Advances in Biosensors: Reviews

Volume 2

S. Yurish
Editor

Advances in Biosensors: Reviews

Volume 2

International Frequency Sensor Association Publishing

S. Yurish, *Editor*
Advances in Biosensors: Reviews. Volume 2

Published by IFSA Publishing, S. L., 2018
E-mail (for print book orders and customer service enquires):
ifsa.books@sensorsportal.com

Visit our Home Page on http://www.sensorsportal.com

ISBN: 978-84-09-05394-0
e-ISBN: 978-84-09-05393-3
BN-20180930-XX
BIC: TCBS

Contents

Contents 5
Preface 9
Contributors 11

**1. Photoelectrochemical Biosensors Boosted
by Nanostructured Materials** 15
  1.1. Introduction 15
  1.2. Working Principles and Potential Applications.................................... 17
    1.2.1. Working Principles *17*
    1.2.2. Potential Applications... *18*
  1.3. Photoactive Species.. 22
    1.3.1. Formation of Composite Semiconductors *24*
    1.3.2. Organic Dye Sensitization.. *26*
    1.3.3. Metal Ion Doping... *28*
  1.4. Sensing Mechanisms of Photoelectrochemical Biosensors 29
    1.4.1. Direct Charge Transfer... *30*
    1.4.2. Enzymatic Reaction.. *31*
    1.4.3. Steric Hindrance *33*
  1.5. Conclusions and Perspective 34
  References... 35

**2. Point-of-Care Electrochemical Immunosensors Applied
to Diagnostic in Health** 43
  2.1. Introduction 43
  2.2. PoCT Electrochemical Immunosensor 46
  2.3. Advances on PoCT Immunosensors.. 47
    2.3.1. PoCT Immunosensors Based on Carbon Allotropes *47*
    2.3.2. PoCT Immunosensors Based on Metal Nanoparticles *51*
  2.4. Lab-On-A-Chip Based Immunosensors 54
  2.5. Conclusions.. 57
  Acknowledgments.. 57
  References... 58

**3. Sensors and Bioelectronics in the Kidney Replacement
Therapy Applications** 65
  List of Abbreviations.. 65
  3.1. Introduction 66
  3.2. Bioimpedance and Impedance Spectroscopy Methods in Biomedical
    Applications.. 69

3.2.1. Water Volumes Evaluation in Bioimpedance Technique *69*
3.2.2. On the Theory of Interfacial Polarization *71*
3.2.3. Multifrequency Bioelectrical Impedance /Spectroscopy (MF-BIA/BIS).
Bioimpedance Analysis in Water Content Measurements *74*
3.2.4. Challenges in the Theory of Multifrequency Electrical
Impedance Analysis *77*
3.3. The Acoustic and Impedimetric Methods in Biomedical Applications 78
3.3.1. Pre-Treatment of the QCM Surface in Biosensors' Applications *81*
3.3.1.1. Supported Membranes and Biomimetic Strategies 81
3.3.1.2. Lipid Vesicles Adhesion: Modelling Cell- Substrate Interaction 84
3.3.2. The Cells Adhesion to the Surfaces Measured with the QCM *85*
3.3.3. QCM for the Studies of Clotting: Heparin Coating, Platelets
Activation and Immunity Factors *89*
3.3.4. QCM for the Whole Blood Studies and Anticoagulants Research *90*
3.3.5. QCM for the Control of Hemodialysis Membranes Biofouling *92*
3.3.6. Theoretical Background for the Acoustic Biosensors (QCM
and SAW-Based): Modelling the Response of Sensors
in Measurements of Soft Biological Materials in Liquids *93*
3.3.7. QCM-EIS/ECIS Combined Sensing *100*
3.3.8. QCM, ECIS or QCM Combined with EIS/ECIS in Water Control
for Dialysis *103*
3.3.8.1. Importance of Water Quality Control for HD 103
3.3.8.2. Designing Biosensors for the Detection of Heavy Metals Ions
in Solution 104
3.3.8.3. Biosensors for Label-Free Sensing of Water Contaminants
(Based on QCM and EIS/ECIS) 107
3.4. Outlook 109
Acknowledgements 112
References 112

4. Solid State Colorimetric Biosensors **133**
4.1. Introduction 133
4.2. Polymer Films and Glass Substrates 134
4.3. Gels 136
4.4. Molecularly Imprinted Polymers 137
4.5. Nanofibers 138
4.6. Other Substrates 141
4.7. Conclusions 141
References 142

5. Sensing of H_2O_2 as Cancer Biomarker with LDHs Nanostructures **147**
5.1. Hydrogen Peroxide as Cancer Biomarker 147
5.2. LDHs Based Electrochemical Biosensors 148
5.3. Fabrication of Nanohybrids 150
5.3.1. Simple Coprecipitation Method *150*
5.4. Cell Culture 152
5.5. Amperometric Detection of Extracellular H_2O_2 152
5.6. Electrocatalytic Sensing Performance 152
5.7. Detection of H_2O_2 Triggered by Live Cancer Cells 155

5.8. Detection of H_2O_2 in Human Serum and Urine Samples 159
5.9. Conclusions 160
References 161

6. Highly Reliable Metallization on Polymer and Their Fundamental Characteristics Toward Wearable Devices Applications 165
6.1. Introduction 165
6.2. Nickel-Phosphorus Metallization 167
  6.2.1. Nickel-Phosphorus Metallization on Polyimide *167*
    6.2.1.1. Sc-CO_2 Catalyzation and CONV Metallization 167
    6.2.1.2. Effects of Sc-CO_2 Catalyzation in Metallization 173
    6.2.1.3. Sc-CO_2 Catalyzation and Sc-CO_2 Metallization 183
    6.2.1.4. Effects of Sc-CO_2 Catalyzation in Sc-CO_2 Metallization 189
  6.2.2. Nickel-Phosphorus Metallization on Nylon 6,6 .... *201*
  6.2.3. Nickel-Phosphorus Metallization on Silk *208*
6.3. Platinum Metallization 218
  6.3.1. Platinum Metallization on Nylon 6,6 .... *218*
  6.3.2. Platinum Metallization on Silk *221*
    6.3.2.1. Pt(acac)$_2$ Sc-CO_2 Catalyzation and CONV Metallization 221
    6.3.2.2. Pd(acac)$_2$ Sc-CO_2 Catalyzation and CONV Metallization 227
6.4. Conclusions 237
Acknowledgements 237
Reference 238

Index 243

Preface

After successful publication of *'Advances in Biosensors: Reviews'*, Vol. 1, Book Series in 2017, and feedback from our authors and readers, we have decided to publish the second volume in 2018. The 2nd volume is also an open access book and available in both formats: electronic and paper (hardcover).

The global biosensors industry is approaching towards multi-billion market. The modern biosensors market growth is driven by the continuous technological advancements in the biosensors ecosystem, increase in the use of biosensors for nonmedical applications, lucrative growth in point-of-care diagnostics, and rise in the demand for glucose monitoring systems. Increasing applications in diagnosis of various diseases and development of nanoparticle based electrochemical biosensors significantly stimulates growth of biosensors industry.

The second volume of *'Advances in Biosensors: Reviews'*, Book Series contains six chapters written by 24 authors from 7 countries: Brazil, China, Denmark, Japan, South Africa, Sweden and Ukraine.

Chapter 1 focuses on the design and fabrication of photo-electrochemical biosensors and their potential applications in analytical detection of some clinically significant biochemical molecules. The working principles and nanostructured construction of such biosensors for enhanced performance are described. Some striking examples are highlighted. The current-status and critical challenges are summarized, and the outlook of this newly emerged type of biosensors is prospected.

Chapter 2 describes point-of-care electrochemical immunosensors applied to diagnostic in health. Particular attention is given on advanced immunosensors based on carbon allotropes and metal nanoparticles.

Chapter 3 is dedicated to the multifrequency bioimpedance analysis, a non-invasive electrical method for monitoring blood and water volume of patients, and the experimental and theoretical developments of the QCM and QCM-D sensors used for the cells- artificial membranes interactions and in blood coagulation studies.

Chapter 4 reflects on the advances of solid-state colorimetric biosensors. The subsections are divided in terms of the solid support material used.

Chapter 5 describes a sensing of H_2O_2 as cancer biomarker with layered doubled hydroxides nanostructures. A new avenue to design afterward generation of bionanoelectronics and miniaturized biosensors as sensitive cancer detection probe is provided.

Chapter 6 describes highly reliable metallization on polymer and their fundamental characteristics toward wearable devices applications.

I hope that readers will enjoy this new volume and that can be a valuable tool for those who are involved in research and development of different biosensors and biosensing systems.

I am looking for any advices, comments, suggestions and notices from the readers to make the next volumes of becoming popular *'Advances in Biosensors: Reviews'* Book Series.

Sergey Y. Yurish,
Editor IFSA Publishing   *Barcelona, Spain*

Contributors

Muhammad Asif, Key Laboratory of Material Chemistry and Service Failure, School of Chemistry and Chemical Engineering, Huazhong University of Science and Technology, Wuhan, 430074, P. R. China

Ayesha Aziz, Key Laboratory of Material Chemistry and Service Failure, School of Chemistry and Chemical Engineering, Huazhong University of Science and Technology, Wuhan, 430074, P. R. China

Diego G. A. Cabral, Biomedical Engineering Laboratory, Federal University of Pernambuco, Av. Prof. Moraes Rego, 1235, Recife, Brazil

Tso-Fu Mark Chang, Institute of Innovative Research, Tokyo Institute of Technology, Yokohama, 226–8503, Japan
CREST, Japan Science and Technology Agency, Yokohama, 226–8503, Japan

Chun-Yi Chen, Institute of Innovative Research, Tokyo Institute of Technology, Yokohama, 226–8503, Japan
CREST, Japan Science and Technology Agency, Yokohama, 226–8503, Japan

Qijin Chi, Department of Chemistry, Technical University of Denmark, DK-2800 Kongens Lyngby, Denmark, E-mail: cq@kemi.dtu.dk; Tel.: +45 45252032; Fax: +45 45883136, Denmark

Wan-Ting Chiu, Institute of Innovative Research, Tokyo Institute of Technology, Yokohama, 226–8503, Japan
CREST, Japan Science and Technology Agency, Yokohama, 226–8503, Japan

Rosa F. Dutra, Biomedical Engineering Laboratory, Federal University of Pernambuco, Av. Prof. Moraes Rego, 1235, Recife, Brazil
E-mail: rosa.dutra@ufpe.br

Paula A. B. Ferreira, Biomedical Engineering Laboratory, Federal University of Pernambuco, Av. Prof. Moraes Rego, 1235, Recife, Brazil

Leonid Y. Gorelik, Chalmers University of Technology, Gothenburg, Sweden

Tomoko Hashimoto, Department of Clothing Environmental Science, Nara Women's University, Nara, 630–8506 Japan

Hiromichi Kurosu, Department of Clothing Environmental Science, Nara Women's University, Nara, 630–8506 Japan

Hongfang Liu, Key Laboratory of Material Chemistry and Service Failure, School of Chemistry and Chemical Engineering, Huazhong University of Science and Technology, Wuhan, 430074, P. R. China

Nokuthula Ngomane, School of Chemical and Physical Sciences, University of Mpumalanga, Mbombela, 1200, South Africa

Mitsuo Sano, Institute of Innovative Research, Tokyo Institute of Technology, Yokohama, 226–8503, Japan
CREST, Japan Science and Technology Agency, Yokohama, 226–8503, Japan

Gilvânia M. de Santana, Biomedical Engineering Laboratory, Federal University of Pernambuco, Av. Prof. Moraes Rego, 1235, Recife, Brazil

Anne K. S. Silva, Biomedical Engineering Laboratory, Federal University of Pernambuco, Av. Prof. Moraes Rego, 1235, Recife, Brazil

Evgen I. Sokol, Kharkiv National Technical University, KhPI, Kharkiv, Ukraine

Masato Sone, Institute of Innovative Research, Tokyo Institute of Technology, Yokohama, 226–8503, Japan
CREST, Japan Science and Technology Agency, Yokohama, 226–8503, Japan

Erika K. G. Trindade, Biomedical Engineering Laboratory, Federal University of Pernambuco, Av. Prof. Moraes Rego, 1235, Recife, Brazil

Marina V. Voinova, Chalmers University of Technology, Gothenburg, Sweden

Zhengyun Wang, Key Laboratory of Material Chemistry and Service Failure, School of Chemistry and Chemical Engineering, Huazhong University of Science and Technology, Wuhan, 430074, P. R. China

Byung-Hoon Woo, Institute of Innovative Research, Tokyo Institute of Technology, Yokohama, 226–8503, Japan

Yao Wu, Department of Chemistry, Technical University of Denmark, DK-2800 Kongens Lyngby, Denmark

Chapter 1

Photoelectrochemical Biosensors Boosted by Nanostructured Materials

Yao Wu and Qijin Chi

1.1. Introduction

Effective clinical diagnosis largely depends on the sensitive and timely identification of clinically relevant molecules, including DNA, peptides, small molecules and antibodies, as they carry out biologically crucial functions from storing and transmitting genetic information, regulating protein activities, enzymatically catalyzing chemical reactions, to transporting small molecules to their target destinations [1]. Nowadays, a number of well-established and conventional clinical analytical techniques, especially enzyme-linked immunosorbent assay (ELISA), chromatography and mass spectrometry are being challenged by point-of-care testing (POCT) devices, due to obvious shortcomings of conventional techniques such as significant financial cost, limited availability in remote and rural areas [2-3], and time-consuming procedures. With the ever-growing demands for ultrasensitive biomolecular detection, there have long been substantial efforts on developing new bioanalytical methods that could provide imperative solutions to personalized health care (PHC). PHC is being appreciated globally to detect various metabolic or infectious disorders since nowadays clinical analysis can not only be performed in clinical laboratories, but it also can routinely be carried out in several settings, including hospital point-of-care settings, by caregivers in nonhospital settings, and by patients at home. The adoption of PHC has shown significantly improved clinical diagnosis in various conditions, including emergency, ambulatory, and remote areas. In a PHC set-up, biofluids are

Qijin Chi
Department of Chemistry, Technical University of Denmark, Denmark

analyzed using various diagnostic devices including lab based equipment and biosensors [4].

Biosensors are defined as analytical devices, which harness the exquisite sensitivity and specificity of biology in conjunction with physicochemical transducers. Based on specific biological recognition, the simple, easy-to-use and miniaturized biosensors could offer an approach to the growing need for rapid, specific, inexpensive, and fully automated means of biomolecular analysis, ranging from medical diagnostics through drug discovery, food safety, process control and environmental monitoring, to defense and security applications [5]. The basic concept of the biosensor was first elucidated by Leyland C. Clark in 1962 [6]. In a Clark oxygen electrode, the glucose was entrapped by a dialysis membrane; electrochemical detection of oxygen or hydrogen peroxide could be used as the basis for a broad range of bioanalysis by incorporation of appropriate enzymes. Inspired by this pioneering work, the biosensor field has grown enormously and, especially, the enzymatic biosensors have been one of the most frequently applied and successful sensors in various biomedical fields, thanks to the widely applied glucose sensors, which account for $10.71 billion market value and are projected to grow at a compound annual growth rate of 5.4 % to reach $14.68 billion in 2022 [7].

Indeed, electrochemical bioanalysis has been a dominating biosensing tool in the market and has generated huge impacts on the biomedical field. With the prompt development of nanotechnology and materials science, the merge of the photoelectrochemical (PEC) process to electrochemical bioanalysis is creating real opportunities to advance electrochemical techniques, since the PEC technique is also the evolutionary generation of the electrochemical methods [8-10]. Thus, the PEC technique not only inherits the advantages of the electrochemical bioanalysis such as low cost, simple instrumentation and easy miniaturization, but it also provides an elegant route for probing various clinically relevant biomolecules such as DNA, proteins, metabolites and small molecules.

In this chapter, we mainly focus on the design, fabrication, basic principles and sensing strategies of PEC biosensors. *First,* PEC biosensors are discussed in the context of the working principles and their potential applications in analytical detection of some clinically significant biomolecules; *second,* some photoactive species that are proposed to accelerate the electron transfer of PEC reactions are exemplified; *third*, the major sensing strategies of PEC biosensors are categorized; *finally*, we conclude and put forward a perspective on the future development of PEC biosensors. We hope that this up-to-date review would serve as a useful resource to inform and discuss with the interested audience for the recent developments in PEC biosensors.

1.2. Working Principles and Potential Applications

1.2.1. Working Principles

In the past decade, we have witnessed the growing popularity of PEC biosensors in biological sciences [11-12]. Combining the PEC technique with bioanalysis, the PEC biosensing approaches have been developed. Consisting of two critical ingredients, the photoactive species are employed to generate the detection signal, and the biological recognition elements are incorporated to detect analytes of interest. Thus, a typical PEC biosensor could convert specific biocatalytic events into electrical signals via the interaction between the photoactive species and the biocatalytic reaction chain.

General instrumentation and the working principles of PEC bioanalysis are schematically illustrated in Fig. 1.1.

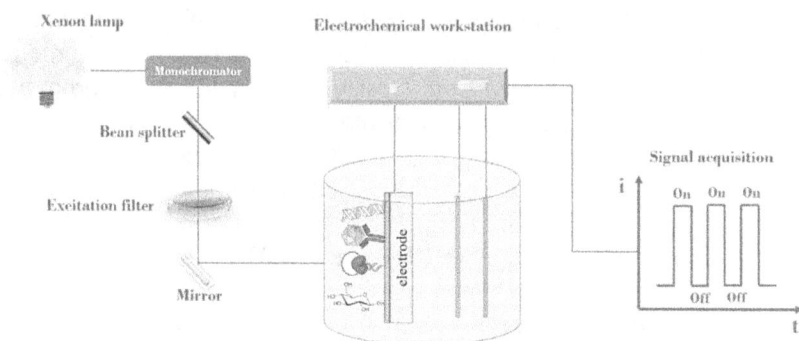

Fig. 1.1. Schematic illustration of the instrumentation and working principles of PEC bioanalysis.

The PEC instrument system generally includes an excitation source (irradiation light, a monochromator, and a chopper), a cell, and an electrochemical workstation with a three-electrode system (working, reference and counter electrodes). Especially, the working electrode comprises of various photoactive species that connected to a metal contact, then to the external electronics, and eventually to a metal counter electrode. Simultaneously, the photoactive species-based working electrode is also connected to the counter electrode via the electrolyte in the electrochemical cell that integrates the circuit. Based on such a system, a specific biorecognition event could be elegantly converted by the photoactive species to the output electrical signals.

1.2.2. Potential Applications

Distinguished form the optical and traditional electrochemical techniques [13-18], the implementation of electronic detection makes PEC instrument cheap, simple and easy to miniaturization while the total separation of excitation source and detection signal enables this method to have low background interference and desirable sensitivity. Thus, it is highly promising that PEC biosensors achieve high performance and enable sensitive detection of various clinically relevant biomolecules such as DNA, proteins, and clinically relevant small molecules (Table 1.1).

DNA is a crucial molecule that carries the genetic instructions used in the growth, development, functioning and reproduction of all known living organisms as well as many viruses [29]. Thus, if an accumulation of genetic and epigenetic changes in DNA such as point mutations, chromosomal rearrangements, microsatellite instability, or promoter hypermethylation occurs, formation and growth of cancerous tumors could be the consequence [30]. As a result, abnormalities in structure or the expression of certain genes can be used to distinguish a cancer cell from a healthy cell. Shown in Fig. 1.2a is a PEC DNA sensor based on hybridization chain reaction (HCR) induced DNA amplification for intercalating photoactive intercalator iridium (III) [31]. When triethanolamine (TEOA) or dissolved O_2 is used as a sacrificial electron donor/acceptor, either cathodic or anodic photocurrent can be generated and used as the readouts for the detection. Under optimal conditions, the PEC DNA sensor displayed photocurrent that is linearly proportional to the logarithm of target DNA concentration in the range from 0.025 to 100 pM with a limit of detection (LOD) down to 9.0 fM.

Table 1.1. Photoelectrochemical Biosensors for the Detection of DNA, Proteins, and Clinically Relevant Small Molecules.

Photoactive Species	Analyte of Interest	Performance	Ref.
TiO$_2$/CdSeTe @CdS:Mn core–shell quantum dots-sensitized TiO$_2$	Carcino-embryonic antigen (CEA, Ag)	Liner range 0.5 pg mL^{-1}–100 ng mL^{-1} with 0.16 pg mL^{-1} as the LOD; Sensitive, reproducible, specific, and stable	[19]
Manganese-doped CdS@ZnS core–shell nanoparticles (Mn:CdS@ZnS)	Caspase-3	Sensitive, reproductive, and stable in studying the nilotinib-induced apoptosis of K562 CML cells	[20]
Carbon quantum dots (CQDs)-functionalized MnO$_2$	TATA binding protein (TBP)	Liner range 2.6 fM–512.8 pM with 1.28 fM as the LOD; Reproducible; Long-term stability	[21]
CdSe/ZnS nanocrystals	Glucose	Dynamic range 100 μM–5 mM; Short response time; Nonstructured sensing electrode	[22]
CdS:Mn@Ru-(bpy)$_2$(dcbpy) nanocomposites	Adenosine triphosphate (ATP)	Liner range 0.5 pM–5 nM with 0.18 pM as the LOD; Highly sensitive, reproducible, and stable	[23]
Graphene–CdS (GR–CdS) nanocomposite	Glutathione (GSH)	Liner range 10 μM–1.5 mM with 3 μM as the LOD; Low applied potential; Rapid response; No interference from anticancer drugs	[24]
IrO$_2$–hemin–TiO$_2$ nanowire	Glutathione (GSH)	Liner range 10 nM–10 μM with 10 nM as the LOD; Low applied potential; Rapid response; Long-term stability; Reproducible	[25]
PbS quantum dots (QDs)	DNA and Thrombin	Liner range 0.8 fM–10 pM with 0.2 fM as the LOD for DNA; Liner range 0.1 pM–10 nM with 15 fM as the LOD for thrombin; Ultrasensitive, versatile, label-free, reproducible and substrate free	[26]
CdTe QDs	DNA	Liner range 1 pM–50 pM with 0.76 pM as the LOD; Rapid, specific, sensitive and reproducible	[27]
SnO$_2$ nanoparticle-modified ITO electrode	DNA	Liner range 100 pM–8 nM with 94 fM as the LOD; Simple, selective, sensitive and generalizable	[28]

Proteins are the biological molecules most ubiquitously affected in diseases [32]. Tens of thousands of core proteins consist of plasma, the contents of which represent secretions coming from all body tissues and may feature markers of physiological and pathological processes. Clinically relevant protein markers include tumor, cardiac, hepatic, inflammatory, and other biomarkers. A new ultrasensitive PEC immunosensing platform based on antibodies conjugated TiO$_2$/CdSeTe@CdS:Mn compete with antibodies conjugated CuS nanocrystals for the antigen is shown in Fig. 1.2b [33]. In this work, carcinoembryonic antigen (CEA), antibodies (Ab1) and signal CEA antibodies (Ab2) are immobilized on the TiO$_2$/CdSeTe@CdS:Mn sensitized structure and CuS nanocrystals, respectively. The detection of CEA is based not only on the competitive absorption of exciting light and consumption of the electron donor between the sensitization structure and CuS nanocrystals, but also on the steric hindrance induced by CuS-Ab2. Because of these dual effects, a low LOD of 0.16 pg/mL and a wide linear range from 0.5 pg/mL to 100 ng/mL for CEA detection were achieved. Thus, this platform has demonstrated great potential for developing various sensitive PEC immunoassays for the detection of disease-related protein markers.

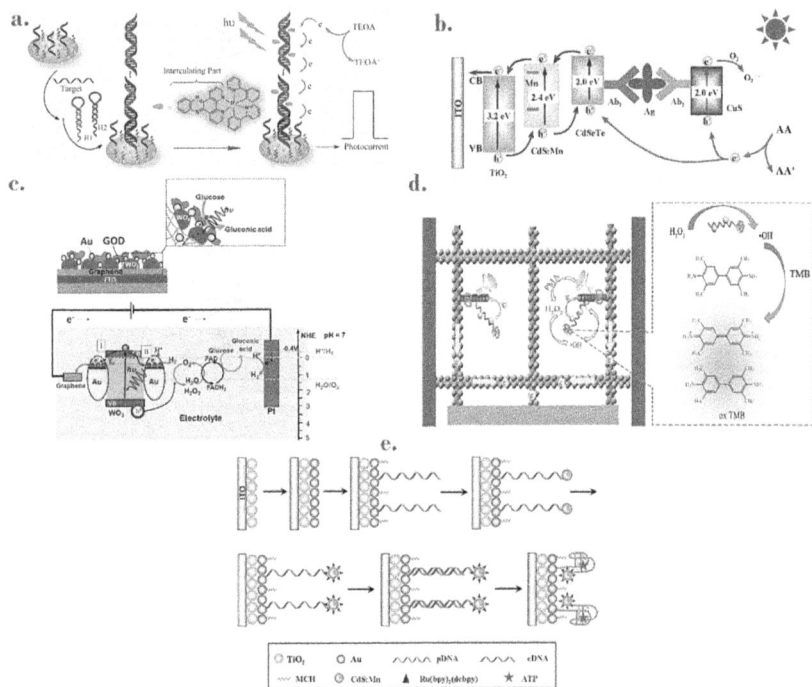

Fig. 1.2. (a) Schematic illustration of the rationale for a PEC DNA sensor. (Reprinted from [31] Copyright 2015 American Chemical Society); (b) Schematic illustration of a new ultrasensitive PEC immunosensing platform for the detection of carcinoembryonic antigen. (Reprinted from [33] Copyright 2016 American Chemical Society); (c) Schematic illustration of graphene-WO$_3$-Au triplet junction for glucose sensing. (Reprinted from [35] Copyright 2014 American Chemical Society); (d) Schematic illustration of the fabrication procedures for the PEC detection mechanism of H$_2$O$_2$ released from MCF-7 Cells. (Reprinted from [37] Copyright 2016 American Chemical Society); (e) Schematic illustration of a novel "signal-on" photoelectrochemical aptasensor for adenosine triphosphate (ATP) detection. (Reprinted from [23] Copyright 2016 American Chemical Society).

Clinically relevant endogenous small molecules and ions include neurotransmitters, metabolites, vitamins, amino acids, dietary minerals, and other small biomolecules. For example, as an important metabolite, the level of glucose in bloods can be used to indicate the metabolic disorder of diabetes [34]. Devadoss et al. [35] demonstrated the PEC oxidation of glucose using graphene-WO_3-Au hybrid membrane modified electrodes. Shown in Fig. 1.2c is a schematic representation of sensing mechanisms of glucose oxidation at the graphene-WO_3-Au hybrid membrane modified with glucose oxidase (GOx) enzyme. Under light irradiation, the photoactive WO_3 nanoparticles generate electron-hole pairs; these photogenerated holes were scavenged by biological analyte acting as electron donors, resulting in the oxidation of the biomolecules through intermediate reaction by electron acceptor FAD/FADH$_2$. In the presence of glucose, glucose is oxidized to gluconic acid at approximately -0.4 V (vs. NHE), while photogenerated electrons reduce water to form H_2. The generated photocurrent passing through the circuit is thus recorded as the sensing signal. Consequently, the PEC sensor performance exclusively depends on the efficiency of photoactive materials in responsive to glucose oxidation and transduction. Besides metabolites, detection of reactive oxygen species (ROS) in cells and assessment of their dynamic release process is essential for understanding the roles of ROS in cellular physiology [36]. A novel dual PEC/colorimetric cyto-analysis format was first introduced into a microfluidic paper-based analytical device (μ-PAD) for synchronous sensitive and visual detection of H_2O_2 released from tumor cells based on an in situ hydroxyl radicals (•OH) cleaving DNA approach, as shown in Fig. 1.2d [37]. The μ-PAD was constructed by a layer-by-layer modification of concanavalin A, graphene quantum dots (GQDs) labeled with flower-like Au@Pd alloy nanoparticles (NPs) probe, and tumor cells on the surface of the vertically aligned bamboo like ZnO, which grows on a pyknotic PtNPs modified paper working electrode (ZnO/Pt-PWE). Thanks to the effective matching of energy levels between GQDs and ZnO, it leads to the enhancement of the photocurrent response compared with the bare ZnO/Pt-PWE. After releasing H_2O_2, the DNA strand was cleaved by •OH generated under the synergistic catalysis of GQDs and Au@Pd alloy NPs and thus, reduced the photocurrent, resulting in a high sensitivity to H_2O_2 in aqueous solutions with a LOD of 0.05 nM. The disengaged probe can result in catalytic chromogenic reaction of substrates, enabling real-time imaging of H_2O_2 biological processes. As a result, this work offers a truly low-cost, simple, and disposable μ-PAD for precise and visual detection of cellular H_2O_2, which shows promising utility to cellular biology and pathophysiology. As the major energy currency of cells, adenosine triphosphate (ATP) plays a crucial role in the regulation of cellular metabolism and biochemical pathways [38]. Thus, the determination of ATP is essential in biochemical studies and clinical analyses. Fan et al. reported a novel "signal-on" PEC aptasensor for the detection of ATP, as shown in Fig. 1.2e [23]. In the assays, a TiO_2/Au hybrid material was used as the PEC matrix for the immobilization of ATP aptamer probes. CdS:Mn@Ru-(bpy)2(dcbpy) photoactive nanocomposite was incorporated as a signal amplification element by labeling on the terminal of ATP aptamer probes. The presence of ATP induces the conformation change of ATP aptamer probes by forming a G-quadruplex structure, resulting in noticeably enhanced photocurrent intensity due to full activation of the sensitization effect. A wide linear range from 0.5 pM to 5 nM with a low detection limit of 0.18 pM was obtained for the ATP detection. Besides these merits including high sensitivity, simplicity, specificity, reproducibility and stability, this sensing platform could further be expanded to detect other clinically important biomolecules.

1.3. Photoactive Species

Inspired by mimicking the energy conversion theme of plants, the photoactive species-based PEC conversion is an attractive approach for converting chemical energy to electricity under light illumination and applied potential [39-40]. With the advances of nanotechnology and materials science, a series of new photoactive species have been proposed and promoted the further development of PEC based biosensors [41-43]. Since the PEC reaction involves charge transfer between electron donor/acceptor and photoactive species at electrode upon light irradiation [44], the choice of photoactive species thus proves to be of paramount importance for effective outcomes in PEC applications.

As one of the most widely applied photoactive species, semiconductor quantum dots (QDs) have been intensively studied especially II-VI semiconductors (e.g. CdSe, CdS, HgS, ZnS, ZnSe, etc.), which can be used for PEC biosensing since an electron-hole pair is generated in the conduction and valence band when QDs are illuminated [45]. Depending on the applied potential, an anodic photocurrent can be generated by electron transfer from the conduction band of the QDs to the electrode, while a cathodic photocurrent can be generated by the electron transfer from the electrode to the valence band of the QDs. As shown in Fig. 1.3, Tanne et al. [22] demonstrated the oxygen sensitivity of electrodes modified with CdSe/ZnS QDs. Illuminating the QDs electrode under polarization, photocurrent can be generated due to the absorption properties of QDs (Fig. 1.3a). During this process, oxygen dissolved in air-saturated buffer is reduced by the generated electrons, resulting in a larger photocurrent (Fig. 1.3b; curve b) than the photocurrent generated in argon-purged buffer (Fig. 1.3b; curve a). Thus, the QDs provide a catalytic layer for the oxygen reduction under illumination, which paves the way for the combination of QDs modified electrodes with the enzymatic activity of GOx in solution. With the presence of the biocatalyst GOx, the QDs based glucose sensor can oxidize glucose by the reduction of O_2, leading to the suppression of the photocurrent of QDs electrode. The QDs electrodes can thus be used for sensing glucose due to the oxygen conversion during the enzymatic reaction process. Indeed, this study offered a general pathway for PEC detection of multianalytes using nanostructured sensing electrodes and localized enzymes. Following this work, more PEC glucose sensors were demonstrated by applying various QDs, such as near-infrared (NIR) QDs [46] and graphene-QDs hybrid systems [47].

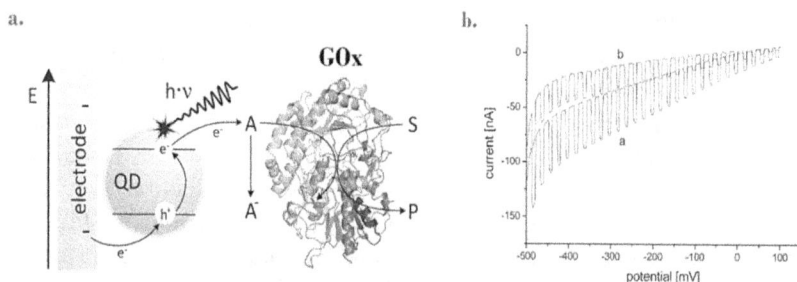

Fig. 1.3. (a) Schematic illustration of the electron transfer steps with the presence of glucose oxidase after illumination of the QD electrode; (b) Linear scan voltammograms of the QD electrode in argon-purged buffer (a) and in air-saturated buffer (b). (Reprinted from [22] Copyright 2011 American Chemical Society).

Bringing in the possibility of driving the PEC bioanalysis with solar light, semiconducting metal oxide prophesies future renewable biosensors since it offers wide band gap and excellent thermal, chemical and structural features. Despite the popularity of semiconductors, practical applications of pure semiconductors are limited to great degree by the rapid recombination of electron-hole pairs. To alleviate this restriction, some sensitization approaches of photoactive species have been proposed to reduce the charge recombination rates and to boost the photocatalytic efficiency. In this section, we focus on discussing three major strategies that have recently been developed to improve the PEC sensing performance: use of composite semiconductors, organic dye sensitization, and metal ions doping. These strategies have demonstrated the ability to accelerate charge transfer in PEC reactions.

1.3.1. Formation of Composite Semiconductors

To date, titanium oxide (TiO_2) has most widely been used in PEC bioanalysis compared with other metal oxides such as tungsten trioxide (WO_3), tin oxide (SnO_2), or zinc oxide (ZnO) [8], due to its unique characteristics including high catalytic efficiency, excellent photochemical stability and biocompatibility, good corrosion resistance and environmental benignity. In recent years, PEC detection of biomolecules such as DNA, glucose, NADH etc. are successfully demonstrated using TiO_2 nanomaterials [48-51]. Despite the preliminary successes, the direct utilization of pure TiO_2 nanomaterials for biosensing applications is still limited because of the wide band gap nature of TiO_2 [52], which absorbs UV light that could deactivate target biomolecules [53]. Consequently, functionalizing TiO_2 surfaces with photoabsorbers that share matched energy levels could offer favorable charge separation to the composite semiconductors [54]. The IrO_2-hemin-TiO_2 nanowires (NWs) for the GSH detection is a good example, as schematically shown in Fig. 1.4a. Hemin, a synthetic molecule that mimics the natural enzyme, was attached to the surface of TiO_2 for analyzing the glutathione (GSH) level with rapid response and good reproducibility [25]. In this IrO_2-hemin-TiO_2 NWs based biosensing for the detection of GSH, hemin recognizes the presence of GSH, which acts as the electron donor to the holes located on the excited state of hemin and the catalyst IrO_2 can facilitate the hole transfer process. Afterwards, these electrons were then quickly injected into the conduction band of the TiO_2 NWs, thanks to the higher oxidation potential of the excited state of hemin than the conduction band energy level of TiO_2 NWs. Finally, electrons dwelling on the conduction band of TiO_2 NWs were subsequently transferred to the fluorine-doped tin oxide (FTO) electrode. Thus, a label-free sensitive, selective and real-time PEC detection of GSH in buffer and cell extracts become feasible by the IrO_2-hemin-TiO_2 NWs composite, suggesting a great potential using composite semiconductors for analyzing the GSH level in various biological samples.

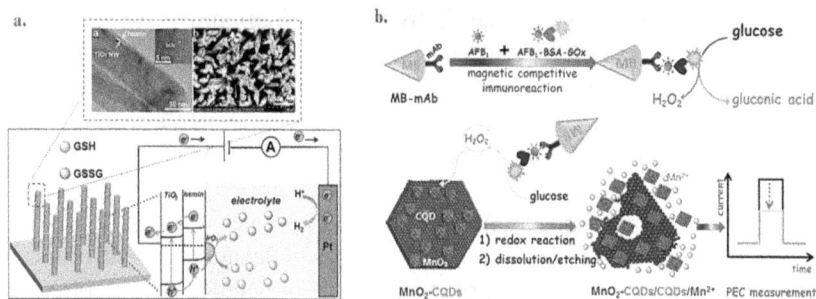

Fig. 1.4. (a) Schematic illustration of the PEC process for GSH detection by an IrO_2–hemin–TiO_2 NW biosensor (inset: TEM and SEM images of TiO_2 NWs). (Reprinted from [25] Copyright 2013 American Chemical Society); (b) Schematic illustration of PEC immunosensing platform toward aflatoxin B_1 (AFB1) on carbon quantum dots-coated MnO_2 nanosheets (MnO_2-CQDs) by coupling an enzyme immunoassay format with a magneto-controlled microfluidic device. (Reprinted from [57] Copyright 2017 American Chemical Society).

Surely, the photoactive material of PEC sensors mainly depends on QDs and metal oxide-based nanostructures. Nonetheless, the potential of QDs and metal oxide-based nanostructures is diminished by drawbacks such as strong oxidation characteristics causing biomolecule damage, requirement for a high-energy excitation optical source, and instability caused by photocorrosion. Encouragingly, carbon-based nanomaterials such as fullerenes (C_{60}) and hybrid nanostructures may open a new horizon for the improvement of PEC properties such as the electron transfer efficiency and the suppression of the charge recombination probability [55-56]. Very recently, Lin et al. [57] designed a hybrid nanostructure with improved PEC activity by coupling manganese oxide (MnO_2) with carbon QDs (CQDs). As shown in Fig. 1.4b, a composite semiconductor with improved PEC activity is synthesized for quantitative monitoring of Aflatoxin B_1 (AFB1, as a model mycotoxin) by coupling with a semi-automatic microfluidic device. Upon addition of target AFB1, the analyte initially competes with AFB1-bovine serum albumin-GOx (AFB1-BSA-GOx) for the labeled anti-AFB antibody on magnetic bead; the carried GOx then oxidizes the glucose to generate H_2O_2, which can be used as an efficient scavenger for the decomposition of MnO_2 nanosheets. Because of the decomposition of MnO_2-CQDs on the electrode, the photocurrent of MnO_2-CQDs modified electrode decreases with the increasing H_2O_2 concentration. By monitoring the change in photocurrent, quantitative determination of AFB_1 in the sample was achieved. Benefited from the hybrid nature, not only the charge-separation rate of MnO_2 can be improved by the CQDs, but the light intensity on the substrate surface also would be increased, resulting from the small-size CQDs and highly transparent MnO_2 nanosheets. Thus, the developed composite semiconductor exhibited ultrahigh sensitivity, good reproducibility and acceptable accuracy for the detection of target AFB_1. As a proof-of-concept design, this platform can also be used for the sensing of other small molecules or mycotoxins, thereby representing a versatile immunosensing scheme.

1.3.2. Organic Dye Sensitization

Although QDs incorporation and dyes sensitization have been investigated to improve the photocurrent conversion efficiency of the PEC biosensors [58-59], the weak adsorption of dyes on the semiconductor layers limited their applications in developing highly sensitive biosensors. To mitigate such limitation, localized surface plasmon resonance (LSPR) of AuNPs have been applied since the LSPR of AuNPs can not only act as light harvesting antennae for dye that improves the light harvest ability, but it also enhances the charge transfer efficiency, which makes it promising in the development of PEC biosensors with high performances [60]. Shown in Fig. 1.5a is a novel visible-light PEC biosensor based on the LSPR enhancement and dye sensitization that was fabricated for highly sensitive analysis of protein kinase activity with ultralow background [61]. In this strategy, DNA conjugated gold nanoparticles (DNA@AuNPs) were assembled on the phosphorylated kemptide modified TiO_2/ITO electrode through the chelation between Zr^{4+} ions and phosphate groups, then followed by the intercalation of $[Ru(bpy)_3]^{2+}$ into DNA grooves. The adsorbed $[Ru(bpy)_3]^{2+}$ can harvest visible light to produce excited electrons that inject into the TiO_2 conduction band to generate photocurrent under visible light irradiation. Thanks to the excellent conductivity and large surface area of AuNPs that facilitate electron-transfer and accommodate large number of $[Ru(bpy)_3]^{2+}$ ions, the photocurrent was significantly amplified, affording an extremely sensitive PEC analysis of kinase activity with ultralow background signals. Thus, the developed dye-sensitization and LSPR enhancement visible-light PEC biosensor shows great potential in protein kinases-related clinical diagnosis and drug discovery.

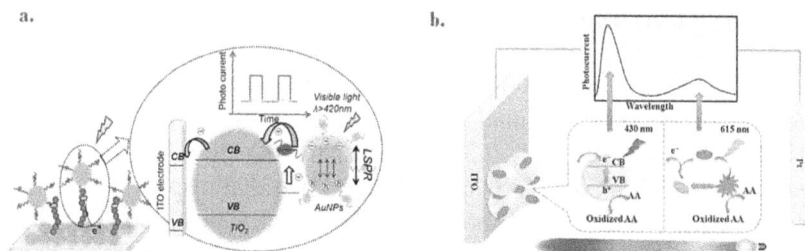

Fig. 1.5. (a) Schematic illustration of the Photoelectrochemical Biosensor for Kinase Activity Detection. (Reprinted from [61] Copyright 2016 American Chemical Society); (b) Schematic illustration of the WR-PEC photocurrent generation process of CdS and MB in a visible light range, and the sensing principle for Cu^{2+}. (Reprinted from [62] Copyright 2016 The Royal Society of Chemistry).

Conventionally, PEC biosensors have been based on a single wavelength of light or a white light source using one photoactive species. As only one output signal could be provided by most of the PEC biosensors, the sensitivity, anti-interference ability and the detection performance for complex samples are compromised. A PEC sensor with two absorption-separated photoactive species that could simultaneously output two PEC signals is thus desirable, since better sensitivity, anti-interference ability and the detection performance could be expected. Hao et al. reported another means to apply dye sensitization in a PEC biosensor, as shown in Fig. 1.5b [62]. In such a setup, a wavelength-resolved ratiometric PEC (WR-PEC) sensor with better anti-interference ability was constructed by a stepwise assembly of CdS QDs and methylene blue (MB) on an ITO electrode. The detection mechanism relies on the quenching effect of Cu^{2+} towards CdS QDs and the photocurrent enhancement by MB through catalytic oxidation of leuco-MB. Consequently, a wavelength-resolved ratiometric approach was realized, and the I_{pQD}/I_{pMB} exhibited a wide linear relationship with the logarithmic value of the Cu^{2+} concentration with high sensitivity. Besides sensitivity, compared with conventional PEC approaches, the WR-PEC sensor showed excellent capacity for anti-interference in complex environments or to experimental option variation. More significantly, the ratiometric sensors were also shown to expand the application of the WR-PEC technique. As a result, the WR-PEC technique provides a novel concept for the design of ratiometric PEC sensors and demonstrates promising applications in biosensing and clinical diagnosis.

1.3.3. Metal Ion Doping

CdS is a widely used semiconductor with a band gap of 2.4 eV corresponding to the optimal absorption range of middle-wavelength visible light (<520 nm) [63]. Thanks to the lifetime of charge recombination for CdS:Mn is much longer than that of pure CdS, Mn^{2+} are introduced into CdS to form CdS:Mn doped nanostructure, so that charge recombination could be depressed. Due to their different band gaps, various semiconductors have their own optimal absorption. To enhance the light absorption efficiency, promote the electron transfer, and prolong the lifetime of charge carriers, a new ultrasensitive PEC immunosensing platform based on antibodies conjugated $TiO_2/CdSeTe@CdS:Mn$ compete with antibodies conjugated CuS nanocrystals for the antigen is shown in Fig. 1.6a [33]. In their proposal, carcinoembryonic antigen (CEA) antibodies (Ab1) and signal CEA antibodies (Ab2) are immobilized on the $TiO_2/CdSeTe@CdS:Mn$ sensitization structure and CuS nanocrystals, respectively. The detection of CEA is based not only on the competitive absorption of exciting light and consumption of the electron donor between the sensitization structure and CuS nanocrystals, but also on the steric hindrance induced by CuS-Ab2. Thanks to the exhibited good specificity, reproducibility and stability, the platform has great potential for developing various sensitive PEC immunoassays for the detection of disease-related biomarkers.

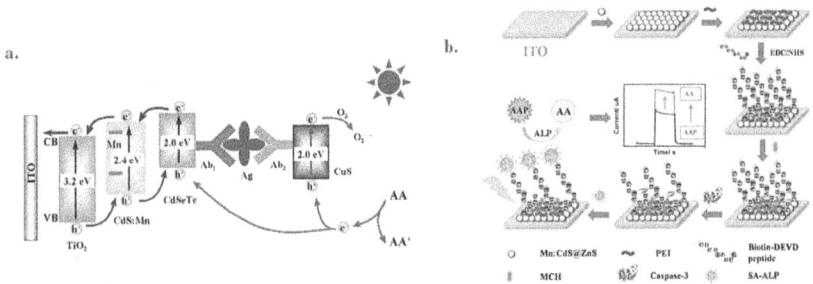

Fig. 1.6. (a) Schematic illustration of a metal ion doped photoelectrochemical immunosensing platform for the detection of carcinoembryonic antigen (CEA, Ag). (Reprinted from [33] Copyright 2016 American Chemical Society); (b) Schematic illustration of a metal ion doped photoelectrochemical sensing platform for antileukemia drug evaluation. (Reprinted from [20] Copyright 2014 American Chemical Society).

Mn-doped CdS@ZnS core-shell nanoparticles (Mn:CdS@ZnS) were used as the photoactive material for antileukemia drug evaluation (Fig. 1.6b) [20]. The Mn:CdS@ZnS nanoparticles generated higher and more stable photocurrent signals than that by pure CdS nanoparticles due to several reasons. *First*, the metal ion dopant, Mn^{2+}, effectively created an electronic state in the midgap region of CdS that prompted charge separation, inhibited recombination dynamics, and broadened the wavelength range of light absorption, thus increasing the photocurrent signal [64]. *Second*, the ZnS shell, which served as a passivation layer, effectively increased the photocurrent of CdS by prohibiting the formation of surface defects on CdS and by inhibiting charge recombination with the external redox couple [65]. *Third*, ZnS also increased the photocurrent signal due to its absorption of ultraviolet light. Through amide bonding, a biotin-Gly-Asp-Gly-Asp-Glu-Val-Asp-Gly-Cys (biotin-DEVD) peptide is immobilized on the nanoparticles, which interacts with streptavidin-labelled alkaline phosphatise (SA-ALP). With the presence of the substrate 2-phospho-lascorbic acid (AAP), the product of the ALP catalyzed reaction, ascorbic acid (AA) is produced, which induces the enhanced photocurrent due to its efficient electron donating capability. Since caspase-3 cleaves the DEVD-biotin peptide, the activity of caspase-3 could thus be detected by monitoring the diminished photocurrent. Because caspase-3 is one of the most frequently activated cysteine proteases and is a well-established marker of cellular apoptosis, the catalytic activity of caspase-3 could thus be used to evaluate the therapeutic effects of anticancer drugs. The efficacy of nilotinib for inducing apoptosis of the K562 CML cells were demonstrated by studying the catalytic activity of caspase-3, indicating that this platform is sensitive, reproductive and stable in studying the nilotinib-induced apoptosis of K562 CML cells, and might be promising for evaluating other anticancer drugs.

1.4. Sensing Mechanisms of Photoelectrochemical Biosensors

The applications of PEC bioanalysis are very dynamic, with increasing explorations in the development of various ingenious detection protocols with novel signaling strategies. Generally, the signaling strategies depend closely on how the biocatalytic/bioaffinity events interact with the PEC nanosystems. By utilizing the biorecognition event to affect the redox reaction of photoexcited material with sacrificial species in solution, the most common signaling strategies can be proposed and intensely studied to probe various targets of interest. Their PEC responsive mechanisms can be classified into the three main categories: *direct charge transfer, enzymatic reaction for the consumption/ generation of electron donor/acceptor,* and *steric hindrance caused by biomolecular recognition reaction.*

1.4.1. Direct Charge Transfer

Among all the PEC methods, using the direct redox reactions of the QDs toward biochemical species in solution might be the simplest PEC method for biomolecular detection. Based on this strategy, the photocurrent changes could be obtained as the signal readout, resulting from the direct communication between the photogenerated charge carriers and the solution-solubilized electron donor or acceptor. A number of PEC bioanalytical studies relied on direct charge transfer have been established for PEC detection of various biomolecules such as glucose [22], cysteine [66], and glutathione [67].

As shown in Fig. 1.7a, a PEC biosensor based on direct charge transfer was designed for the detection of cysteine using Au–SnO$_2$/CdS hybrid nanospheres (HNSs) [66]. Under visible-light irradiation, cysteine scavenged the hole of the excited CdS as an electron donor, followed by the photoinduced electron transfer in CdS from its valence band (VB) to its conduction band (CB), and then electrons injected to the CB of SnO$_2$. In this system, AuNPs served as an electron relay to mediate the electron transfer from SnO$_2$ to the ITO electrode, thus resulting in efficient overall charge transport and an enhanced photocurrent signal. Due to the superior electrical conductivity of Au, AuNPs serves as an excellent electron-transport matrix to capture electrons and transport electrons from excited CdS to ITO rapidly, avoiding electron–hole recombination, resulting in a much more sensitive response to visible light. This strategy displays advantages such as rapid response, wide detection range, low cost, good reproducibility, and promise to be used in the fields of catalysis, PEC devices and biosensors.

Another example of direct charge transfer based PEC biosensor is shown in Fig. 1.7b [67]. CdS/RGO/ZnO nanowire array heterostructure was synthesized for innovative self-powered PEC detection of glutathione. Under irradiation, both ZnO and CdS can be excited to generate electron–hole pairs. Due to the type II alignment between ZnO and CdS, the photogenerated electron–hole pairs can be efficiently separated [68]. The excited electrons on the CB of CdS are favorably injected to the CB of ZnO, which are subsequently transferred to the FTO substrate, while the excited holes on the VB of ZnO flow favorably to the VB of CdS. Besides, RGO can facilitate the transfer process of the excited electrons on the CB of CdS to ZnO because of its superior charge collection and shuttle characteristics [69]. The self-powered device demonstrates satisfactory sensing performance with rapid response, a wide detection range, an acceptable detection limit, as well as certain selectivity, reproducibility, stability, which is promising for the development of PEC biosensors.

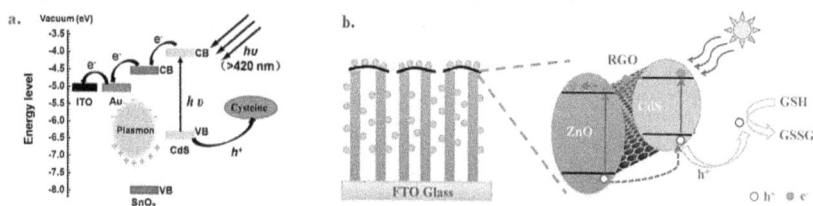

Fig. 1.7. (a) Schematic illustration of the electron-transfer process at the Au–SnO$_2$/CdS/ITO electrode under visible light. (Reprinted from [66] Copyright 2014 The Royal Society of Chemistry); (b) Schematic illustration of the mechanism for the PEC detection of GSH. (Reprinted from [67] Copyright 2015 Wiley-VCH Verlag GmbH & Co. KGaA, Weinheim).

1.4.2. Enzymatic Reaction

Complementary to the direct charge-transfer mechanism, another straightforward approach is to use enzymatic reaction for the consumption/generation of electron donor/acceptor. Among them, dissolved O$_2$, ascorbic acid (AA) and H$_2$O$_2$ serving as the reactants can react with photoactive species to generate stable photocurrent. A variety of enzymes such as GOx, sarcosine oxidase (SOD), acetylcholine esterase (AChE), alkaline phosphatase (ALP), horseradish peroxidase (HRP), β-galactosidase (β-Gal), glucose dehydrogenase (GDH), fructose dehydrogenase (FDH), alcohol dehydrogenase (ADH), and DNAzyme etc. could be employed as biocatalysts to consume the soluble substrates and/or produce soluble products that acts as electron donor/acceptor species to trigger the photocurrent variation [70-71].

For example, an enzyme-based PEC biosensor for the detection of ethanol is shown in Fig. 1.8a [72]. In this proposal, a nanocomposite

31

containing graphene nanosheets and CdS QDs (GNs-CdS QDs) were modified onto a glassy carbon (GC) electrode for the electrochemical oxidation of chlorpromazine (CPZ) to chlorpromazine-sulfoxide (CPZ-SO), which simultaneously oxidized NADH at reduced overpotential. Since the photogenerated electrons in QDs have more time to be transferred to graphene while photogenerated holes oxidize reduced form of CPZ-SO which is in direct contact with QDs, electron-hole pairs recombination was suppressed, and the observed photocurrent was enhanced significantly. By immobilizing ADH onto the modified electrode via a simple cross-linking procedure, the proposed system could be used for ethanol biosensing. Due to that the different mediators can be used as electron donor in this system, the designed strategy could be converted to versatile PEC platforms for various PEC biosensing applications.

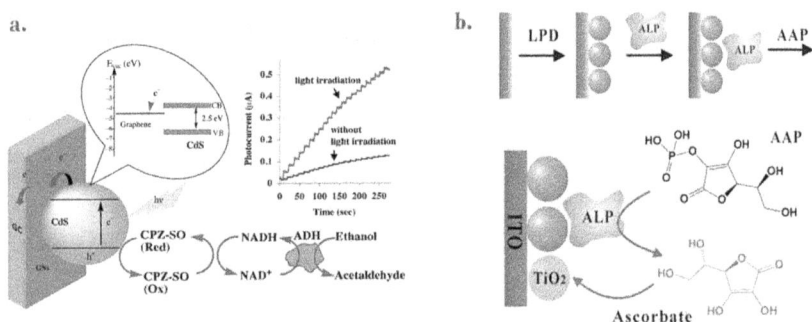

Fig. 1.8. (a) Schematic illustration of charge separation and the enhanced electron transfer in GNs-CdS QDs/IL/CPZ-SO modified GC electrode and photoelectrocatalytic oxidation of NADH. The electrochemical and photoelectrochemical biosensing process of ethanol is also illustrated. (Reprinted from [72] Copyright 2014 Wiley-VCH Verlag GmbH & Co. KGaA, Weinheim); (b) Schematic illustration of the proposed PEC assembly consisting of ALP catalyzed transformation of AAP to ascorbate for in situ activating the inert TiO₂ nanocrystallites. (Reprinted from [73] Copyright 2013 American Chemical Society).

Using the ALP catalytic process to couple with the unique surface chemistry of nanocrystalline TiO$_2$, as shown in Fig. 1.8b, a strategy for PEC enzymatic analysis has also been developed for enzyme activity or inhibitor detection [73]. ALP was immobilized onto the film of TiO$_2$ NPs with the ITO glass as the transparent back contact to allow the catalyzed hydrolysis of ascorbic acid 2-phosphate (AAP). The dephosphorylation process generated the product of ascorbate containing enediol ligands, which have high affinity to the undercoordinated surface Ti atoms on TiO_2 NPs [74]. Due to the in-situ self-coordination of these produced surface-active ligands onto the nanosized TiO_2 surface, the latter would experience a natural adjustment in the coordination geometry, endowing the inert semiconductor with strong absorption bands in the visible region, and hence underlying a novel, simple, sensitive and general PEC bioanalysis strategy.

1.4.3. Steric Hindrance

Distinguished by its effective diffusion suppression of electron donor/acceptor to photoactive species surface, steric hindrance caused by molecular recognition reaction is another signaling format for designing PEC biosensors. The common molecular recognition reactions with the specific binding interaction of antibody-antigen, biotin-avidin, DNA-protein, aptamer-protein, etc. can all be used [75-77].

For example, by means of antibody-antigen interaction, Yang et al. developed an immunosensing platform for PEC detection of Ochratoxin A (OTA) as shown in Fig. 1.9b [77]. A general label-free PEC platform was manufactured by assembly of CdSe NPs sensitized anatase TiO_2-functionalized electrode via layer-by-layer (LBL) strategy. CdSe NPs were assembled on anatase TiO_2-functionalized electrode through dentate binding of TiO_2 NPs to –COOH groups. Under visible-light irradiation, AA was used as an efficient electron donor for scavenging photogenerated holes. As a result of the band alignment of CdSe and TiO_2 in electrolyte, the photocurrent response of the CdSe NPs modified electrode was significantly enhanced. Antibodies of OTA were immobilized on CdSe sensitized electrode by using the classic 1-ethyl-3-(3-dimethylaminopropyl) carbodiimide hydrochloride coupling reactions between –COOH groups on the surfaces of CdSe NPs and –NH_2 groups of the antibody. This PEC platform established a simple, fast and inexpensive strategy for fabrication of label-free biosensor, which could be generalizable for other types of applications.

Likewise, Ma et al. developed another immunosensing platform for PEC detection of TATA binding protein (TBP) using DNA-protein interaction, as shown in Fig. 1.9b [21]. Through the complementary pairing between ssDNA fragments, AuNPs were brought into the intimate distance of the CdS QDs modified onto the ITO electrode for the inter-particle interactions. Specifically, the binding of the TBP would bend the AuNPs capped dsDNA, thereby placing the Au NPs closer to the CdS QDs on the electrode. With changed inter-particle distance, the inter-particle ET efficiency would be influenced, and the quenching effect of AuNPs on the CdS QDs could be strengthened accordingly. Besides, by using the dsDNA sequence as a rigid spacer the relationship between the photocurrent intensity and the spacing distance was able to be studied via varying the length of dsDNA. This work thus presents a novel mechanism for ET-based PEC protein assays with high sensitivity and selectivity.

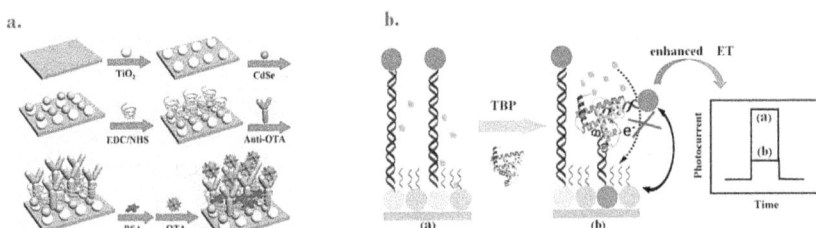

Fig. 1.9. (a) Schematic illustration of the PEC immunosensor for the detection of Ochratoxin A. (Reprinted from [77] Copyright 2015 Elsevier); (b) Schematic illustration of the PEC immunosensor for the detection of TATA binding protein. (Reprinted from [21] Copyright 2016 American Chemical Society).

1.5. Conclusions and Perspective

Photoelectrochemical bioanalysis provides a crucial analytical tool for sensitive and selective determination of analytes of interest. Meanwhile, unlike spectroscopic and chromatographic instruments, PEC bioanalysis can be easily adapted for detecting a wide spectrum of biological species. In recent years, numerous novel PEC systems have been developed and new protocols have been tested. In this book chapter, we present recent progress in the applications of PEC biosensors by focusing on their design, fabrication, basic principles, wildly applied photoactive species and major signaling strategies. Due to the continued development in photoactive species, customized biomolecular systems, and their ingenious incorporation into PEC bioanalytical platforms, research in this field has continued boosting towards more sophisticated systems. Tremendous achievements have been made in PEC biosensors so far: by drawing on the progress and accomplishment of materials science, many functionalized and nontoxic nanomaterials with high PEC activity and stability have been employed in PEC biosensing; by incorporating a recognition unit (antibodies, protein receptors, or peptide aptamers) into the multifunctional nanomaterials with excellent biocompatibility, new platforms for various applications have been developed; by taking full advantage of novel manufacturing technologies, such as 3D printing and microelectromechanical systems, miniaturization and even ultrafine integrated processes and batch manufacturing with negligible differences could be implemented; by incorporating automatic technology into microarray platforms, accurate and high-throughput assays for large quantities of complex samples might be feasible.

However, to further expand the use of PEC biosensors and enhance detection performance of PEC biosensors, some aspects should be explored for further work: 1) Fabrication of new and nontoxic materials with high photoelectricity activity in the PEC bioanalysis for achieving superior synergy effects; 2) Exploitation of feasible protocols capable of simultaneous and/or multiplex determination; 3) Integration of PEC bioanalysis into paper-based microfluidic systems to realize robust, portable, or miniaturized devices for particular applications; 4) Employment of the systems for measurements in complex biological matrixes like urine, saliva, blood, etc., and resolving the concomitant issues of fouling and stability; and 5) Development of more reliable detection systems and their application for practical purposes.

Along with the constant developments in science and technology, great opportunities and bright prospects for the development of PEC biosensors could be foreseen. The advancement of PEC biosensors would certainly lead to significant advantages compared to the current systems in terms of simplicity, speediness, cost, and automation. We would like to believe that in the near future, advanced PEC bioanalysis would be feasible for simultaneously monitoring multiple targets in real samples with single miniaturized analyzers.

References

[1]. L. Hood, Systems biology and new technologies enable predictive and preventative medicine, *Science*, Vol. 306, 2004, pp. 640-643.
[2]. K. E. McCracken, J.-Y. Yoon, Recent approaches for optical smartphone sensing in resource-limited settings: a brief review, *Analytical Methods*, Vol. 8, 2016, pp. 6591-6601.

[3]. N. A. Meredith, C. Quinn, D. M. Cate, T. H. Reilly, J. Volckens, C. S. Henry, Paper-based analytical devices for environmental analysis, *Analyst*, Vol. 141, 2016, pp. 1874-1887.

[4]. K. Mahato, A. Srivastava, P. Chandra, Paper based diagnostics for personalized health care: Emerging technologies and commercial aspects, *Biosensors and Bioelectronics*, Vol. 96, 2017, pp. 246-259.

[5]. A. P. F. Turner, Biosensors: Sense and sensibility, *Chemical Society Reviews*, Vol. 42, 2013, pp. 3184-3196.

[6]. C. Clark, L. C. Lyons, Electrode systems for continuous monitoring in cardiovascular surgery, *Annals New York Academy of Sciences*, Vol. 102, 1962, pp. 29-45.

[7]. Frost & Sullivan, Future of Diabetes Care Paradigms – Forecast to 2022, https://www.researchandmarkets.com/reports/4211634/future-of-diabetes-care-paradigms-forecast-to#pos-6

[8]. A. Devadoss, P. Sudhagar, C. Terashima, K. Nakata, A. Fujishima, Photoelectrochemical biosensors: New insights into promising photoelectrodes and signal amplification strategies, *Journal of Photochemistry and Photobiology C: Photochemistry Reviews*, Vol. 24, 2015, pp. 43-63.

[9]. W.-W. Zhao, J.-J. Xu, H.-Y. Chen, Photoelectrochemical bioanalysis: The state of the art, *Chemical Society Reviews*, Vol. 44, 2015, pp. 729-741.

[10]. W. W. Zhao, J. J. Xu, H. Y. Chen, Photoelectrochemical enzymatic biosensors, *Biosensors and Bioelectronics*, Vol. 92, 2017, pp. 294-304.

[11]. M. Liang, L.-H. Guo, Photoelectrochemical DNA sensor for the rapid detection of DNA damage induced by styrene oxide and the Fenton reaction, *Environmental Science and Technology*, Vol. 41, 2007, pp. 658-664.

[12]. Y. Zang, J. Lei, H. Ju, Principles and applications of photoelectrochemical sensing strategies based on biofunctionalized nanostructures, *Biosensors and Bioelectronics*, Vol. 96, 2017, pp. 8-16.

[13]. Y. T. Long, C. Kong, D. W. Li, Y. Li, S. Chowdhury, H. Tian, Ultrasensitive determination of cysteine based on the photocurrent of nafion-functionalized CdS-MV quantum dots on an ITO electrode, *Small*, Vol. 7, 2011, pp. 1624-1628.

[14]. S. J. Xie, H. Zhou, D. Liu, G. L. Shen, R. Yu, Z. S. Wu, In situ amplification signaling-based autonomous aptameric machine for the sensitive fluorescence detection of cocaine, *Biosensors and Bioelectronics*, Vol. 44, 2013, pp. 1-6.

[15]. Y. Yuan, X. Gan, Y. Chai, R. Yuan, A novel electrochemiluminescence aptasensor based on in situ generated proline and matrix polyamidoamine dendrimers as coreactants for signal amplication, *Biosensors and Bioelectronics*, Vol. 55, 2014, pp. 313-317.

[16]. P. Kumar, P. Ramulu Lambadi, N. Kumar Navani, Non-enzymatic detection of urea using unmodified gold nanoparticles based aptasensor, *Biosensors and Bioelectronics*, Vol. 72, 2015, pp. 340-347.

[17]. H. Zhu, G. C. Fan, E. S. Abdel-Halim, J. R. Zhang, J. J. Zhu, Ultrasensitive photoelectrochemical immunoassay for CA19-9 detection based on CdSe@ZnS quantum dots sensitized TiO_2NWs/Au hybrid structure amplified by quenching effect of Ab2@V^{2+}conjugates, *Biosensors and Bioelectronics*, Vol. 77, 2016, pp. 339-346.

[18]. Y. Liu, H. Ma, Y. Zhang, X. Pang, D. Fan, D. Wu, Q. Wei, Visible light photoelectrochemical aptasensor for adenosine detection based on CdS/PPy/g-C_3N_4 nanocomposites, *Biosensors and Bioelectronics*, Vol. 86, 2016, pp. 439-445.

[19]. G. C. Fan, H. Zhu, D. Du, J. R. Zhang, J. J. Zhu, Y. Lin, Enhanced photoelectrochemical immunosensing platform based on CdSeTe@CdS:Mn core-shell quantum dots-sensitized TiO_2 amplified by CuS Nanocrystals conjugated signal antibodies, *Analytical Chemistry*, Vol. 88, 2016, pp. 3392-3399.

[20]. S. Zhou, Y. Kong, Q. Shen, X. Ren, J. R. Zhang, J. J. Zhu, Chronic myeloid leukemia drug evaluation using a multisignal amplified photoelectrochemical sensing platform, *Analytical Chemistry*, Vol. 86, 2014, pp. 11680-11689.

[21]. Z. Y. Ma, Y. F. Ruan, F. Xu, W. W. Zhao, J. J. Xu, H. Y. Chen, Protein binding bends the gold nanoparticle capped DNA sequence: Toward novel energy-transfer-based photoelectrochemical protein detection, *Analytical Chemistry*, Vol. 88, 2016, pp. 3864-3871.

[22]. J. Tanne, D. Schäfer, W. Khalid, W. J. Parak, F. Lisdat, Light-controlled bioelectrochemical sensor based on CdSe/ZnS quantum dots, *Analytical Chemistry*, Vol. 83, 2011, pp. 7778-7785.

[23]. G. C. Fan, M. Zhao, H. Zhu, J. J. Shi, J. R. Zhang, J. J. Zhu, Signal-on photoelectrochemical aptasensor for adenosine triphosphate detection based on sensitization effect of CdS:Mn@Ru(bpy)2(dcbpy) nanocomposites, *Journal of Physical Chemistry C*, 120, 2016, pp. 15657-15665.

[24]. X. Zhao, S. Zhou, Q. Shen, L.-P. Jiang, J.-J. Zhu, Fabrication of glutathione photoelectrochemical biosensor using graphene-CdS nanocomposites, *Analyst*, Vol. 137, 2012, pp. 3697-3703.

[25]. J. Tang, B. Kong, Y. Wang, M. Xu, Y. Wang, H. Wu, G. Zheng, Photoelectrochemical detection of glutathione by IrO_2-hemin-TiO_2 nanowire arrays, *Nano Letters*, Vol. 13, 2013, pp. 5350-5354.

[26]. G. L. Wang, J. X. Shu, Y. M. Dong, X. M. Wu, W. W. Zhao, J. J. Xu, H. Y. Chen, Using G-quadruplex/hemin to 'switch-on' the cathodic photocurrent of p-type PBS quantum dots: Toward a versatile platform for photoelectrochemical aptasensing, *Analytical Chemistry*, Vol. 87, 2015, pp. 2892-2900.

[27]. G. Wen, X. Yang, Y. Wang, Quantum material accompanied nonenzymatic cascade amplification for ultrasensitive photoelectrochemical DNA sensing, *Journal of Material Chemistry B*, Vol. 5, 2017, pp. 7775-7780.

[28]. X. Zhang, Y. Xu, Y. Zhao, W. Song, A new photoelectrochemical biosensors based on DNA conformational changes and isothermal circular strand-displacement polymerization reaction, *Biosensors and Bioelectronics*, Vol. 39, 2013, pp. 338-341.

[29]. A. Travers, G. Muskhelishvili, DNA structure and function, *Federation of European Biochemical Societies Journal*, Vol. 282, 2015, pp. 2279-2295.

[30]. L. Wu, X. Qu, Cancer biomarker detection: recent achievements and challenges, *Chemical Society Reviews*, Vol. 44, 2015, pp. 2963-2997.

[31]. C. Li, H. Wang, J. Shen, B. Tang, Cyclometalated iridium complex-based label-free photoelectrochemical biosensor for DNA detection by hybridization chain reaction amplification, *Analytical Chemistry*, Vol. 87, 2015, pp. 4283-4291.

[32]. N. L. Anderson, N. G. Anderson, The human plasma proteome, *Molecular and Cellar Proteomics*, Vol. 1, 2002, pp. 845-867.

[33]. G. C. Fan, H. Zhu, D. Du, J. R. Zhang, J. J. Zhu, Y. Lin, Enhanced Photoelectrochemical immunosensing platform based on CdSeTe@CdS:Mn Core-Shell quantum dots-sensitized TiO_2 amplified by CuS Nanocrystals conjugated signal antibodies, *Analytical Chemistry*, Vol. 88, 2016, pp. 3392-3399.

[34]. J. Wang, Electrochemical glucose biosensors, *Chemical Reviews*, Vol. 108, 2008, pp. 814-825.

[35]. A. Devadoss, P. Sudhagar, S. Das, S. Y. Lee, C. Terashima, K. Nakata, A. Fujishima, W. Choi, Y. S. Kang, U. Paik, Synergistic metal-metal oxide nanoparticles supported electrocatalytic graphene for improved photoelectrochemical glucose oxidation, *ACS Applied Materials and Interfaces*, Vol. 6, 2014, pp. 4864-4871.

[36]. D. Trachootham, J. Alexandre, P. Huang, Targeting cancer cells by ROS-mediated mechanisms: A radical therapeutic approach?, *Nature Review Drug Discovery*, Vol. 8, 2009, pp. 579-591.

[37]. L. Li, Y. Zhang, L. Zhang, S. Ge, H. Liu, N. Ren, M. Yan, J. Yu, Paper-based device for colorimetric and photoelectrochemical quantification of the flux of H_2O_2 releasing from MCF-7 cancer cells, *Analytical Chemistry*, Vol. 88, 2016, pp. 5369-5377.

[38]. E. Llaudet, S. Hatz, M. Droniou, N. Dale, Microelectrode biosensor for real-time measurement of ATP in biological tissue, *Analytical Chemistry*, Vol. 77, 2005, pp. 3267-3273.

[39]. M. G. Walter, E. L. Warren, J. R. McKone, S. W. Boettcher, Q. Mi, E. A. Santori, N. S. Lewis, Solar water splitting cells, *Chemical Reviews*, Vol. 110, 2010, pp. 6446-6473.

[40]. Y. Yu, X. Wang, Semiconductor nanowires for energy conversion, *Chemical Reviews*, Vol. 110, 2010, pp. 527-546.

[41]. D. J. K. Swainsbury, V. M. Friebe, R. N. Frese, M. R. Jones, Evaluation of a biohybrid photoelectrochemical cell employing the purple bacterial reaction centre as a biosensor for herbicides, *Biosensors and Bioelectronics*, Vol. 58, 2014, pp. 172-178.

[42]. J. Zhang, L. Tu, S. Zhao, G. Liu, Y. Wang, Y. Wang, Z. Yue, Fluorescent gold nanoclusters based photoelectrochemical sensors for detection of H_2O_2 and glucose, *Biosensors and Bioelectronics*, Vol. 67, 2015, pp. 296-302.

[43]. S. Wu, H. Huang, M. Shang, C. Du, Y. Wu, W. Song, High visible light sensitive MoS2 ultrathin nanosheets for photoelectrochemical biosensing, *Biosensors and Bioelectronics*, Vol. 92, 2017, pp. 646-653.

[44]. R. Gill, M. Zayats, I. Willner, Semiconductor quantum dots for bioanalysis, *Angewandte Chemie International Edition*, Vol. 47, 2008, pp. 7602-7625.

[45]. S. Chen, L. W. Wang, Thermodynamic oxidation and reduction potentials of photocatalytic semiconductors in aqueous solution, *Chemistry Materials*, Vol. 24, 2012, pp. 3659-3666.

[46]. W. Wang, L. Bao, J. Lei, W. Tu, H. Ju, Visible light induced photoelectrochemical biosensing based on oxygen-sensitive quantum dots, *Analytica Chimica Acta*, Vol. 744, 2012, pp. 33-38.

[47]. X. T. Zheng, A. Than, A. Ananthanaraya, D. H. Kim, P. Chen, Graphene quantum dots as universal fluorophores and their use in revealing regulated trafficking of insulin receptors in adipocytes, *ACS Nano*, Vol. 7, 2013, pp. 6278-6286.

[48]. M. Zheng, Y. Cui, X. Li, S. Liu, Z. Tang, Photoelectrochemical sensing of glucose based on quantum dot and enzyme nanocomposites, *Journal of Electroanalytical Chemistry*, Vol. 656, 2011, pp. 167-173.

[49]. Y. Hu, Z. Xue, H. He, R. Ai, X. Liu, X. Lu, Photoelectrochemical sensing for hydroquinone based on porphyrin-functionalized Au nanoparticles on graphene, *Biosensors and Bioelectronics*, Vol. 47, 2013, pp. 45-49.

[50]. G. L. Wang, J. J. Xu, H. Y. Chen, Dopamine sensitized nanoporous TiO_2 film on electrodes: Photoelectrochemical sensing of NADH under visible irradiation, *Biosensors and Bioelectronics*, Vol. 24, 2009, pp. 2494-2498.

[51]. K. Wang, J. Wu, Q. Liu, Y. Jin, J. Yan, J. Cai, Ultrasensitive photoelectrochemical sensing of nicotinamide adenine dinucleotide based on graphene- TiO_2 nanohybrids under visible irradiation, *Analytica Chimica Acta*, Vol. 745, 2012, pp. 131-136.

[52]. P. Sudhagar, K. Asokan, E. Ito, Y. S. Kang, N-Ion-implanted TiO_2 photoanodes in quantum dot-sensitized solar cells, *Nanoscale*, Vol. 4, 2012, pp. 2416-2422.

[53]. Q. Kang, L. Yang, Y. Chen, S. Luo, L. Wen, Q. Cai, S. Yao, Photoelectrochemical detection of pentachlorophenol with a multiple hybrid CdSexTe1-x/TiO2 nanotube structure-based label-free immunosensor, *Analytical Chemistry*, Vol. 82, 2010, pp. 9749-9754.

[54]. P. Liu, X. Huo, Y. Tang, J. Xu, X. Liu, D. K. Y. Wong, A TiO_2 nanosheet-g-C_3N_4 composite photoelectrochemical enzyme biosensor excitable by visible irradiation, *Analytica Chimica Acta*, Vol. 984, 2017, pp. 86-95.

[55]. W. Zhan, K. Jiang, M. D. Smith, H. E. Bostic, M. D. Best, M. L. Auad, J. V. Ruppel, C. Kim, X. P. Zhang, Photocurrent generation from

porphyrin/fullerene complexes assembled in a tethered lipid bilayer, *Langmuir*, Vol. 26, 2010, pp. 15671-15679.

[56]. W. Zhan, K. Jiang, A modular photocurrent generation system based on phospholipid-assembled fullerenes, *Langmuir*, Vol. 24, 2008, pp. 13258-13261.

[57]. Y. Lin, Q. Zhou, D. Tang, R. Niessner, D. Knopp, Signal-on photoelectrochemical immunoassay for aflatoxin B1 based on enzymatic product-etching MnO_2 nanosheets for dissociation of carbon dots, *Analytical Chemistry*, Vol. 89, 2017, pp. 5637-5645.

[58]. Z.-Y. Ma, J.-B. Pan, C.-Y. Lu, W.-W. Zhao, J.-J. Xu, H.-Y. Chen, Folding-based photoelectrochemical biosensor: binding-induced conformation change of a quantum dot-tagged DNA probe for mercury(II) detection, *Chemical Communications*, Vol. 50, 2014, pp. 12088-12090.

[59]. S. Wu, H. Song, J. Song, C. He, J. Ni, Y. Zhao, X. Wang, Development of triphenylamine functional dye for selective photoelectrochemical sensing of cysteine, *Analytical Chemistry*, Vol. 86, 2014, pp. 5922-5928.

[60]. T. Kawawaki, Y. Takahashi, T. Tatsuma, Enhancement of dye-sensitized photocurrents by gold nanoparticles: Effects of plasmon coupling, *Journal of Physical Chemistry C*, 117, 2013, pp. 5901-5907.

[61]. Z. Yan, Z. Wang, Z. Miao, Y. Liu, Dye-sensitized and localized surface plasmon resonance enhanced visible-light photoelectrochemical biosensors for highly sensitive analysis of protein kinase activity, *Analytical Chemistry*, Vol. 88, 2016, pp. 922-929.

[62]. Q. Hao, X. Shan, J. Lei, Y. Zang, Q. Yang, H. Ju, A wavelength-resolved ratiometric photoelectrochemical technique: design and sensing applications, *Chemical Science*, Vol. 7, 2016, pp. 774-780.

[63]. Y. L. Lee, C. F. Chi, S. Y. Liau, CdS/CdSe Co-Sensitized TiO_2 photoelectrode for efficient hydrogen generation in a photoelectrochemical cell, *Chemistry Materials*, Vol. 22, 2010, pp. 922-927.

[64]. P. K. Santra, P. V. Kamat, Mn-doped quantum dot sensitized solar cells: A strategy to boost efficiency over 5 %, *Journal of American Chemical Society*, Vol. 134, 2012, pp. 2508-2511.

[65]. R. Freeman, J. Girsh, I. Willner, Nucleic acid/quantum dots (QDs) hybrid systems for optical and photoelectrochemical sensing, *ACS Applied Materials and Interfaces*, Vol. 5, 2013, pp. 2815-2834.

[66]. Q. Shen, J. Jiang, S. Liu, L. Han, X. Fan, M. Fan, Q. Fan, L. Wang, W. Huang, Facile synthesis of Au-SnO_2 hybrid nanospheres with enhanced photoelectrochemical biosensing performance, *Nanoscale*, Vol. 6, 2014, pp. 6315-6321.

[67]. K. Zhao, X. Yan, Y. Gu, Z. Kang, Z. Bai, S. Cao, Y. Liu, X. Zhang, Y. Zhang, Self-powered photoelectrochemical biosensor based on CdS/RGO/ZnO nanowire array heterostructure, *Small*, Vol. 12, 2016, pp. 245-251.

[68]. R. Marschall, Semiconductor composites: Strategies for enhancing charge carrier separation to improve photocatalytic activity, *Advanced Functional Materials*, Vol. 24, 2014, pp. 2421-2440.

[69]. V. Lightcap, T. H. Kosel, P. V. Kamat, Anchoring semiconductor and metal nanoparticles on a two-dimensional catalyst mat. storing and shuttling electrons with reduced graphene oxide, *Nano Letters*, Vol. 10, 2010, pp. 577-583.

[70]. M. Riedel, N. Sabir, F. W. Scheller, W. J. Parak, F. Lisdat, Connecting quantum dots with enzymes: mediator-based approaches for the light-directed read-out of glucose and fructose oxidation, *Nanoscale*, Vol. 9, 2017, pp. 2814-2823.

[71]. N. Tiwari, V. Vij, K. C. Kemp, K. S. Kim, Engineered carbon-nanomaterial-based electrochemical sensors for biomolecules, *ACS Nano*, Vol. 10, 2016, pp. 46-80.

[72]. F. Jafari, A. Salimi, A. Navaee, Electrochemical and photoelectrochemical sensing of NADH and ethanol based on immobilization of electrogenerated chlorpromazine sulfoxide onto graphene-CdS quantum dot/ionic liquid nanocomposite, *Electroanalysis*, Vol. 26, 2014, pp. 530-540.

[73]. W. W. Zhao, Z. Y. Ma, J. J. Xu, H. Y. Chen, In situ modification of a semiconductor surface by an enzymatic process: A general strategy for photoelectrochemical bioanalysis, *Analytical Chemistry*, Vol. 85, 2013, pp. 8503-8506.

[74]. N. M. Dimitrijevic, Z. V. Saponjic, B. M. Rabatic, T. Rajh, Assembly and charge transfer in hybrid TiO_2 architectures using biotin-avidin as a connector, *Journal of American Chemical Society*, Vol. 127, 2005, pp. 1344-1345.

[75]. N. Haddour, J. Chauvin, C. Gondran, S. Cosnier, Photoelectrochemical immunosensor for label-free detection and quantification of anti-cholera toxin antibody, *Journal of American Chemical Society*, Vol. 128, 2006, pp. 9693-9698.

[76]. D. Fan, D. Wu, J. Cui, Y. Chen, H. Ma, Y. Liu, Q. Wei, B. Du, An ultrasensitive label-free immunosensor based on CdS sensitized Fe-TiO_2 with high visible-light photoelectrochemical activity, *Biosensors and Bioelectronics*, Vol. 74, 2015, pp. 843-848.

[77]. Yang, P. Gao, Y. Liu, R. Li, H. Ma, B. Du, Q. Wei, Label-free photoelectrochemical immunosensor for sensitive detection of Ochratoxin A, *Biosensors and Bioelectronics*, Vol. 64, 2015, pp. 13-18.

Chapter 2

Point-of-Care Electrochemical Immunosensors Applied to Diagnostic in Health

Diego G. A. Cabral, Gilvânia M. de Santana, Paula A. B. Ferreira, Anne K. S. Silva, Erika K. G. Trindade and Rosa F. Dutra

2.1. Introduction

Point-of-care testing (PoCT) have been recognized as one of the most attractive methods for decentralization of analytical practices, being mainly developed to diagnostics that require rapid interventions, such as cardiovascular diseases, drug intoxication, emergency preparedness in surgical procedures, containment of transmissibility, spread of infectious diseases, and surveys in endemic or epidemic outbreaks [1]. Another interesting application of PoCTs devices is in continuous monitoring of markers that require recurrent evaluations, glycemic or mostly in therapies and prolonged treatments of diseases like the cancer, being also benefit for treatment and monitoring of patients that live in areas far from central laboratories. In these situations, the conventional laboratorial testing become impracticable, since samples should be transported, processed and results returned to the doctors. Challenges in developing of PoCTs for medical diagnostics involve to combine the advantages of fast results, low cost and user friendly processing, without loss of diagnostic sensibility and specificity, when they are compared to laboratorial analyses.

PoCTs are analytical devices designed to be used near the bedside, reducing the turnaround time of the diagnostic cycle, usually processed

Rosa F. Dutra
Biomedical Engineering Laboratory, Federal University of Pernambuco, Brazil

outside hospital or laboratory that do not require skilled personnel for managements [2]. Currently, PoCT tests are considered practical and economical methods, being nowadays considered as one of the most attractive analytical possibilities, compared to the chemical analyzers, immunoanalyzers, PCR (polymerase chain reaction) and others [2]. Among PoCT devices, lateral flow assays (LFA) and biosensors addressed to immunoassays are more economically profitable than enzyme-linked immunosorbent assay (ELISA) or Electro-chemiluminescence immunoassay (ECLIA), especially regarding to time of analyzes.

Lateral flow assays (LFA) based on immunochromatographic tests are paper assays that use immobilized antibodies or antigens to capture target analytes in samples. A color band resulted on a paper-strip from molecule (antigen or antibody) or material labeling reveals this reaction, usually supplying qualitative results. However, additional image resources can be used to produce quantitative data based in contrast and brightness of color band [3]. A typical LFA is formed by overlapping membranes mounted on a rigid support, which confers stability and facilitates the handling [4]. Tip of the strip has a sample pad made of adsorbent material, where the sample is applied. The samples are transported by capillarity to the conjugation pad containing the labeled antibodies for biorecognition. The interaction between the target analytes and these antibodies form complexes that migrate to the reaction zone, usually formed by a nitrocellulose membrane. In this zone, there are two lines of immobilized antibodies, one to the target molecule and the control to define the results [8]. A schematic design of a LFA is shown in Fig. 2.1.

On the last few decades, advances of nanotechnology has allowed incorporation of gold nanoparticles to LFA, improving the sensitive of the analytical testing [7] Currently, LFA immunoassay have been developed for several applications, allowing the screening of infectious diseases (HIV, viral hepatitis, tuberculosis and herpes simplex virus and others) [8]. LFA has also been applied to PoCTs of cardiac markers such as troponin, H-FABP, hepatitis and others, possibility a semi-quantitative analysis; however it is quite limited, because the results are color band-dependents, thus the results are subject of human misinterpretation [9].

In attempting to overcome the limitations denoted by LFA, immunosensors can supply a quantifiable signal, as greater as well as higher the analyte concentrations, independently of detected species: antibodies, antigens, enzymes, or other chemical species. The interest for immunosensors has been exponentially growing on the last decades due to combine advantages of high sensitivity, user-friendly processing and portability, beside to present a low cost per analyses (Fig. 2.2).

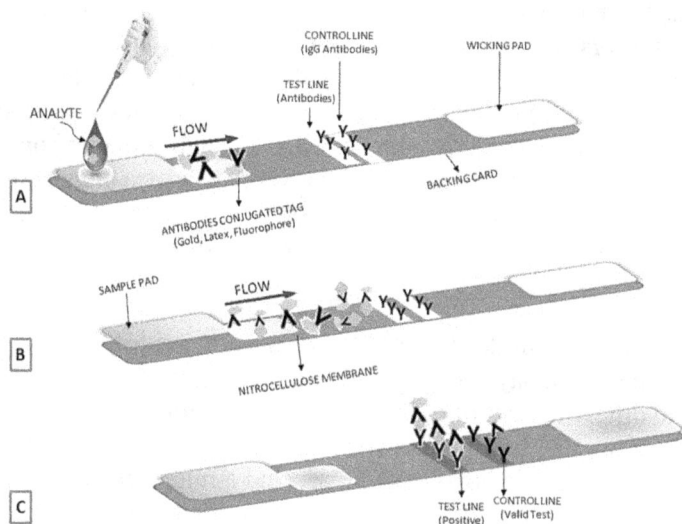

Fig. 2.1. Schematic design of a typical Lateral Flow Immunoassay at different steps: (A) adding the samples containing antigens and immunoglobulins; (B) migration of antigens with labelled antibodies, and (C) immunocomplex formed and immunoglobulis are positioned by affinity on the paper regions where labelled antibodies are exhibited by a color band.

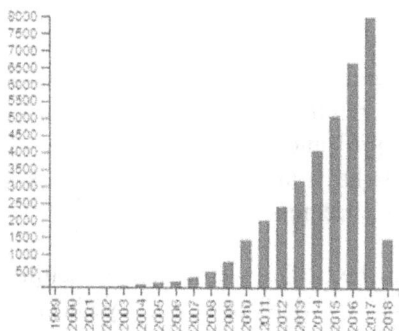

Fig. 2.2. Number of citation over the last decades (Extracted from Web of Science base: ["immunosensor" OR "electrode" OR "biosensor"] AND ["point-of-care"] in March 2018).

2.2. PoCT Electrochemical Immunosensor

Immunosensors are based on the specific antigen-antibody interactions causing a perturbation on the electrode surface by molecule capturing by immobilized antigens or antibodies; this perturbation is converted into measurable signals by a transducer. In general, signal is amplified, processed and readout in output display [10]. Specificity of immunosensors is mainly dependent of affinity between antigen-antibody. Monoclonal and polyclonal antibodies can be used in immobilization technique developments, nevertheless monoclonal antibodies are more attractive due to recognize only a one epitope of an antigen, being more specific, although commonly have a higher cost [11] Recombinant antigens have been more recently used to produce antibodies with more selectively, in order to recognize only one epitope region.

Screen-printed electrode (SPE) has significantly contributed for PoCT developments. The layer-by-layer printing of commercial or self-made inks onto different types of rigid and flexible substrates. Conventionally, SPE comprises one sensing unit with three printed electrodes, including a working electrode, a counter electrode and a reference electrode. The composition of the inks chosen in the printing process is essential to the selective determination intended for each analysis [12]. Commonly, SPE uses voltammetric techniques, measuring changes on current responses produced by a controlled potential (constant or periodic). Current responses are generated by diffusion of redox species from electrolyte/electrode interface, being proportional to the binding events, i.e. antigens or antibodies captured.

In numerous point-of-care immunosensors using amperometric transduction have been developed for clinical diagnostics, such as for HIV [13], prostate specific antigen (PSA) [14], celiac disease [15], cardiac troponin T [16] and cardiac troponin I [17]. Other transducer types using the SPE have also been described for impedance [18] or capacitance [19] measurements.

Recent advances in the SPE development for clinical diagnostic were obtained with progress derived from synthesis of nanostructured electrode surfaces. Metallic nanoparticles, nanowires, carbon nanotubes, graphene, and their respective nanocomposites have been widely used combining with pastes, or forming film on the working electrodes [20]. Using nanocomposites or nanofilms was possible to increase the amount of immobilized biomolecules on the surface sensor due to greater of working electrode area. Additionally, nanomaterials have also contributed to increase the sensitivity of biosensors, due to their electrical, optics, acoustic and other interesting proprieties, particularly indispensable and individuals of each nanomaterial that are capable to produce devices with more reproducible results and robustness [21].

2.3. Advances on PoCT Immunosensors

Nanomaterials have improved specially the efficiency and reliability of electrochemical PoCT immunosensors, allowing a lower limit of detection in the concentrations of antigens or antibodies present in biologic fluid samples that was not previously possible. Nanomaterials can be defined based on size parameter(s), being under 100 nm sized in, at least, one dimension. Commonly, in nanoscale, these materials present new properties that are not normally observed, when they are in bulk. These alterations are obtained by the quantum effects of size, being especially evident in carbon allotropes and metal nanoparticle [22, 23]. For this reason, it is clear that the progress of bioanalytical assays will rely heavily on innovations in nanotechnology [24, 25].

Several nanomaterials have contributed to electrochemical immunosensor developments, among them metal nanoparticles, metal oxides nanoparticles, carbon nanotubes, graphene, their corresponding nanocomposites and quantum dots (Fig. 2.3) [26].

2.3.1. PoCT Immunosensors Based on Carbon Allotropes

Carbon allotropes contribution in electrochemical immuno-PoCTs has gained prominence due to their small size of the carbon atoms and the number of electrons they can share, allowing the formation of several bonding patterns and stable versatile materials with excellent intrinsic properties such as high electrical conductivity, large surface area, ease of functionalization and biocompatibility [27].

Carbon nanotube

Among the nanostructures synthesized from carbon allotropes should be highlighted the carbon nanotubes (CNT) that were discovered in 1991 by Iijima, enabling interaction with biomolecules for biosensor applications [28]. CNTs can be described as hexagonal arrangements in cylindrical format, held by Van der Waals interactions in the adjacent layers. They promote a rapid electron transfer, increasing the reaction rate of many electroactive species, and then decreasing the electrode response time of the Immuno-PoCTs, thereby achieving high sensitivity with low detection limits [29]. With respect to the structure the CNT can be classified in two forms: single wall nanotubes (SWCNT), formed by a single layer of carbon atoms arranged in a hexagonal way, and multiple wall nanotubes (MWCNT), which consists of multiple layers of carbon atoms arranged in a hexagonal way arranged around a central area. The length of a CNT can range from nanometers to centimeters, but the diameter varies in the order of nanometers, depending on the type of CNT; so it possesses a high aspect ratio, or the length-to-diameter ratio, can be as high as 132,000,000:1, which is unequalled by any other material [30]. Activation or functionalization of CNTs by oxidation treatment introduces chemical functional groups, including alcoholic, carboxylic, aldehydic, ketonic, and esteric oxygenated functional groups [31]. These groups allow a greater interaction between CNTs and biomolecules, enabling an oriented immobilization, as example by the Fc portion of antibodies and by exposing their binding sites or antigenic regions to the specific epitopes (Fig. 2.4) [32].

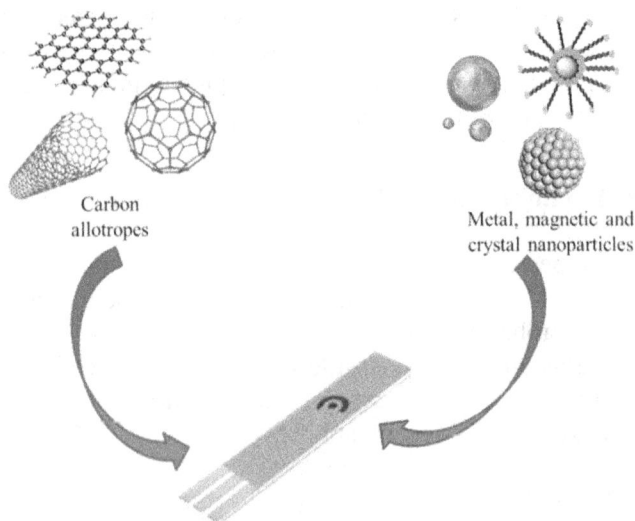

Carbon
allotropes

Metal, magnetic and
crystal nanoparticles

Fig. 2.3. Nanomaterials with potential applications in SPE based-PoCTs. Illustrating of CNTs as carriers or reporters for signal generation and powerful amplifiers for electrochemical transduction.

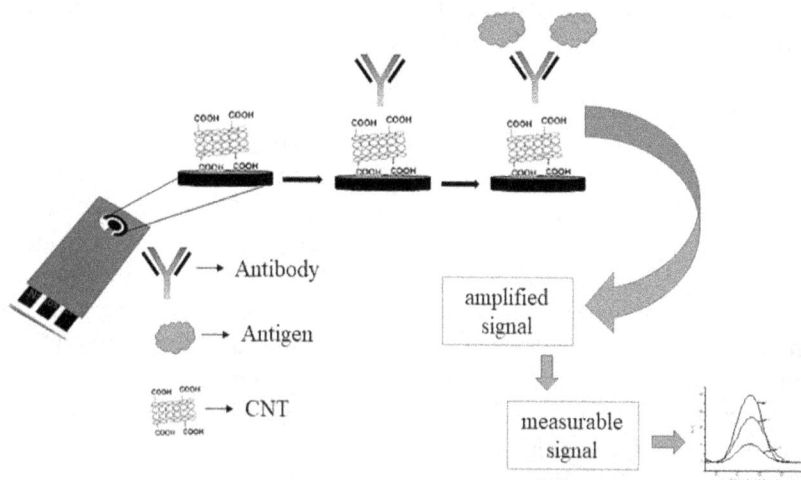

Fig. 2.4. Use of CNTs to detect the interaction between antigen-antibody in PoCT.

Studies have shown that CNTs interact with other materials, improving the intrinsic qualities of the PoCTs immunosensor. Dias et al. [33], (2013) produced a dengue virus for NS1 detection based on a homogeneous mixture consisting of carboxylated CNTs dispersed in carbon ink as a printed working electrode. The effect of the matrix, as well as the performance of the assays, was successfully evaluated using the spiked blood serum sample, obtaining excellent recovery values in the results. The carbon nanotubes incorporated into the carbon paint improved the reproducibility and sensitivity of the CNT-SPE immunosensor. In another work, Silva et al. [34] (2013) developed a label-free immunosensor based on printed electrodes for CNTs functionalized with amine groups to detect the cardiac troponin T. PoCT was developed by the homogenization between the carbon ink and the amine carbon nanotubes forming a thin film on a polyethylene terephthalate substrate. The use of NH_2-CNTs increased the reproducibility and stability of the sensor, and the amine groups allowed an oriented immobilization of antibodies against cardiac troponin T.

Aiming to allow the allignment of CNTs on sensor surface, polymer films have been commonly employed. The polymers interact with the CNT through the functional groups (-COOH, -OH, $-NH_2$) and may form nanocomposites or nanohybrids. As examples we can cite the work of Sanchez-Tirado et al. [35], (2017) where dual screen-printed carbon electrodes modified with 4-carboxyphenyl-functionalized double-walled carbon nanotubes were used for the preparation of electrochemical immunosensors for the simultaneous determination of the cytokines Interleukin-1β (IL-1β) and factor necrosis tumor α (TNF-α). In addition, the dual immunosensor exhibits excellent reproducibility of the measurements, storage stability and selectivity as well as negligible crosstalking. In recent years, studies have shown that the use of CNT combined with conductive polymers can improve sensitivity and increase electron transfer on the sensor platform. In another study Gomes et al. [36], (2013) produced a nanostructured SPE immunosensor based on carbon nanotubes supported by a conductive polymer film for detection of cardiac Troponin T (cTnT). The combined use of polyethyleneimine (PEI) film and CNT provided important advantages for obtaining a highly sensitive analytical method for cTnT.

Graphene

Another prominent carbon nanomaterial is graphene (G). It is a 2D material of atomic thickness, formed by carbon atoms with sp2 hybridization, forming a structure of hexagonal shape similar to a hive. The characterization and identification of graphene was first performed in 2004, after synthesis obtained by successive exfoliation steps of graphite oxide using an adhesive tapes [37]. Among its remarkable properties, it is cited the transparency of the sheet (optical transmittance of ~97.7 %) and large surface area (2630 m^2 g^{-1}). Furthermore, it is a good heat conductor (thermal conductivity of 500 W m^{-1} K^{-1}), chemically inert and a semimetal with high electron transfer (charge mobility of 250 000 cm^2 V^{-1} s^{-1} in room temperature) [38].

According to physical and chemical characteristics, different forms of graphene are shown, among them Graphene Oxide (GO) and Reduced Graphene Oxide (RGO), which, because of their particularities, are highly attractive for the assembly of sensor surfaces [39]. GO has two dimensions, consisting of a hexagonal network of Sp2 bonds between carbon atoms (CC) and by Sp3 bonds with oxygen atoms (CO) forming carboxyl groups (-COOH), hydroxyls (-OH) or epoxy (-O-). This makes GO an excellent material for biological applications, since its functional groups readily interact with nucleic acids, proteins, cells, and other organic molecules [40]. Yukird et al. [41], (2017) developed an electrochemical immunosensor based on a nanohybrid formed by graphene and polyaniline (G/PANI). Electrospraying of G/PANI increased the electrode surface area while electropolymerization of aniline increased the number of amino groups ($-NH_2$) for antibody immobilization.

Reduced graphene oxide (rGO) is another material that has been widely used in electrochemical analysis, due to its high effective surface area and high electrical conductivity [42]. rGO is produced from reduction of GO via thermal, chemical, electrochemical and laser-scribing methods. In rGO synthesis, functional groups are removed and the conductivity is increased again [40]. As examples of the use of graphene for the production of Immuno-PoCTs we can mention the work of Silva et al., (2016) [43], that produced a biomimetic sensor for the detection of Troponin T based on a nanocomposite formed by the conjugation of rGO and Polypyrrole. Another example is Lee et al. [44] (2017), who developed an electrochemical immunosensor for the detection of carcinoembryonic antigen. In this method, silver nanoparticles were mixed with rGO to modify the surface of screen-printed carbon electrode.

2.3.2. PoCT Immunosensors Based on Metal Nanoparticles

Metal nanomaterials have been aroused interest due to their special optical and electrocatalytic properties. They are often incorporated by adhesion or binding to the modified transduction platforms. Robust devices for health applications were fabricated with enhanced performance using nanoparticles [45]. Metal nanoparticles have been used to immobilize antibodies or antigens, and or have served as electronic conduction vehicles in electrochemical biosensors [46]. Metal nanoestructures, semiconductor nanoparticles and metal oxide nanostructures have been considered as potential signal labels when attached to secondary antibodies to stimulate the development of signal amplification strategy for immunosensing [47].

Gold nanoparticles

Gold is an inert metal in macroscale, but gold nanoparticles (GNPs) are adopted nanomaterial often explored as detectable labels to enhance a suitable signal, thereby providing an intense, pronounced and vivid mark. The color change of GNPs are observable with bare eye. This optical property of revelation in visible color is valuable especially in colorimetric assays [48]. Although they have a higher cost, they present high conductivity, excellent biocompatibility, superior stability, low toxicity, relatively simple production and modification [49]. Thus, colloidal GNPs have been used to modify solid electrodes and has shown advantages in feasibly attachment of the immunological molecules and the electron transfer that increase the electrochemical signals. The strong affinity for the amino groups is explored and gold provide a microenvironment compatible with biomolecules, remaining their activity even after immobilization [50]. Moreover, the formation of self-assembled monolayers (SAMs) through oriented Au-S bonds affords great attention to gold toward SPEs for adhesion of more components. Gold SPE helps to deposit antibody in close vicinity with transducer and GNPs help to cast antibody in close vicinity with the antigen and hence results in the increase of sensitivity until femtogram level [51]. Also, Jacobs and coworkers [52] have proposed an 52immunosensors for ultrasensitive detection of troponin-T based on antibody conjugated to GNPs. Using electrochemical impedance spectroscopy, the interdigitated sensor was able to detect concentrations in femtogram per milliliter (fg/mL) of this cardiac marker. Recently, Sabouri et al. [53] have developed a sensitive 52 immunosensors for detection of Hepatitis B virus based on GNPs. HbsAg was targeted by a primary antibody and a secondary antibody co-immobilized on luminol-GNPs, with detection limit of 14 pg/mL.

Silver nanoparticles

Silver is a relatively cheap noble metal that exhibit superior properties over gold on the nanoscale, mainly of optical nature [54]. Its optical profile exhibits the sharpest and most intense bands among metals [55]. Consequently, for convenience, colorimetric assays are prevalent with silver nanoparticles (SNPs) by the straightforward color change discrimination. They can be oxidized more easily and offer improved electrochemical activity, making them good candidates for detection tags in electrochemical sensing. The utilization either naked or conjugated with recognition probes as signal transduction elements for analyte detection in biosensors was shown to improve the detection limits and enhance their diagnostic performance [56]. For this, silver nanostructures need to be associated with recognition molecules that can selectively detect and capture the analyte of interest. However, the functionalization still is a challenging process. They are less stable in aqueous dispersions and are susceptible to oxidation and etching by chloride ions. By their limited stability and difficulty to functionalize, SNPs have become less popular [57]. Considering practical situations,

Hao et al. [58] have developed a direct electrochemical detection approach to assay generically proteins by using SNPs labels coupled covalently with antibody on a SPE. The detection limit found was 0.4 ng/mL. Now, Felici and colleagues [59] have described a novel prototype of label-free 53 immunosensors using SPE and exploiting SNPs as a backing material and electrochemical tracker. Che and coworkers [60] have constructed an amperometric 53 immunosensors for the determination of α-1-fetoprotein, a tumor marker found in several malignant diseases. Multiwalled carbon nanotube-silver nanoparticle composite modified on the surface of a glassy carbon electrode leading a detection limit of 0.08 ng/m. Similarly, Ibupoto and colleagues [61] have described a new potentiometric 53 immunosensors for the selective detection of d-dimer using SNPs decorated ZnO nanotubes anchored to antibodies. D-dimer is a biomarker found at high levels in deep vein thrombosis disorders. It was found a detection limit of 1.00×10^{-6} μg/mL.

Magnetic nanoparticle

Comparatively, magnetic nanoparticles (MNPs) are cheaper to produce, being considered physically and chemically stable, biocompatible and environmentally safe. Magnetic labels have certain peculiarities for biosensing applications, like absence of pre-processing stage for sample purification, since biological entities do not show any magnetic behavior or susceptibility and therefore, no interferences or noise is to expect during signal capturing [62]. Hence, they are also important items for biomedical applications involved in LFA systems as a colored reagent, possessing strong brown coloration. One promising utility is magnetic preconcentration before the detection event. MNPs conjugated to bioreceptor unit can simply be mixed in solution to interact specifically with the analyte. They offer the convenience of separation via external magnetic field, permitting them easily be attracted with a small magnet, losing their magnetic effect when the field is removed. This way, these nanoparticles can be efficiently separated and isolated from the solution [56]. However, the main strategy is the integration of MNPs into the transducer element or the modification of the sensor surface. Despite a wide range of ferromagnetic materials, iron oxides (Fe_2O_3 and Fe_3O_4) are most commonly used for generation or amplification of analytical signal [63]. Employing proper functionalization methods, some notable benefits are achieved such as rapid analysis process, better stability and low detection limit. Besides, they are fluorescent alternatives that offer ease of handling, low production cost and smaller size of final fabricated device when compared to fluorophores [64]. For instance, combining the aforementioned trends, a novel amperometric magnetoimmunoassay based on MNPs pulled by magnetic field on the screen-printed carbon electrodes surface was developed for the selective determination of Legionella pneumophila. The achieved limit of detection by Martín et al. was 104 Colony Forming Units (CFUs)/mL [65]. Singh and Krishnan achieved the first serum insulin voltammetric immunosensor for clinical diagnosis of type 1 and type 2 diabetic disorders. It was reported a lower detection limit of 5 pM for free insulin present in serum using functionalized magnetite nanoparticles [66].

Metal oxide nanoparticles

Zinc oxide (ZnO) also belong group of elite nanomaterials with inherent optical, and piezoelectric properties. It is a semiconducting material that exhibits biomimetic, high catalytic efficiency, little toxicity, low biodegradability, and stable immobilization of proteins due to high isoelectric point without distorting their bioactivity [67]. Beside good electron transfer, this metal oxide nanoparticle denotes a strong adsorption capability, offering numerous sites to antibodies, enzymes and proteins which make them choice for biosensors. It should be conjugated with biological molecules without losing the integrity [68]. For example, a glucose electrochemical sensor based on ZnO nanorods was investigated by Marie and coworkers [69]. The lower limit of detection was 0.22 µM. And a microfluidic immunosensor applied in congenital hypothyroidism screening was presented by Seia and colleagues [70]. ZnO nanobeads were employed as platform for monoclonal antibody immobilization to specifically capture thyrotropin hormone. The electrochemical detection limit of glass microchip was 0.00087 µUI mL.

2.4. Lab-On-A-Chip Based Immunosensors

Due to the in-depth knowledge of nanomaterials, great advances were achieved, making it possible to implement confined labs on a single chip or laboratory analysis system. Lab-on-a-chip combines analysis, reaction and processing in a single microchip, i.e., the ability to gather multiple key functions of a size reduced laboratory on an electronic device with a few square centimeters, which typically manipulates human fluids in the order of microliters to nanoliters [26]. This approach have been extensively applied in point of care devices due to advantages such as compactness, mobility, integrability, modularity, reconfigurability, embedded computing, limited power consumption and minimum need to sample and reagent when enormous amounts of volume are not available [71]. Additionally, lab-on-chip platforms are hermetically enclosed with precise control conditions, avoiding evaporation and minimizing the risk of contamination by potentially infectious biological specimens [72]. Regarding personalized healthcare, the multiple detection by a single PoCT is an important trend which could replace time-consuming laboratory analyses [73]. In addition to releasing results in minutes, they play an important role in management and early investigation of diseases and outbreaks [74]. One of the purposes is the development of a chip-based, miniaturized and portable system that allows for the assay of different analytes in complex samples. In this way, many researches in the scientific community have focused on paper-based and printed electrode technologies as approaches for fabricating these diagnostic systems. These technologies are affordable, user-friendly, rapid, and scalable for manufacturing. Moreover, the association with nanomaterials provides a path for the development of highly sensitive and selective biosensors for prospective generation PoCT tools [21].

Paper-based microfluidics or lab on paper is a novel system for handling and analysis of fluid extracellular for a variety of medical applications, such as healthcare and screening [75]. This technology presents simplicity, portability, disposability, low-cost and allows the automation of multi-step processes [76]. Nitrocellulose membrane, chromatography paper and filter paper are attractive substrates for fabricating microfluidic device, because they are natural, porous, ubiquitous and inexpensive materials. Confining solvents and reagents in specific points, paper can drive and regulate aqueous movement passively using capillary forces without supplying of some kind of external energy, and the migration perform the sorting, mixing and uniform separation of the liquid samples diffused [77]. Furthermore, the chemical composition of paper permits the covalent bonding of bioactive compounds onto the surface. On the other hand, some obstacles to become an ideal PoCT are liquid evaporation, sample retention and nonspecific adsorption. These adversities could lead to false response errors and decreased sensitivity [21]. Its mode of construction is creating a set of microchannels bounded by hydrophobic barriers patterned on paper substrates witch the flow is conducted within the hydrophilic channels and consequently, fluid can be coordinated of a controlled mode. Two-dimensional (2D) and three-dimensional 3D microfluidic channels have been already built on paper, being able to transport biological liquids injected separately by pathways for performing assays and quantifying concentrations of distinct analytes [78]. Printing is the one of the most commonly used techniques to achieve minimal consumption of hydrophobic material [79]. A wave of advancements in 3D printing technology to simplify in agile designing and fabrication supports the durability, flexibility and performance of PoCT microfluidic [80].

Different detection methods have been employed for a semi-quantitative detection, analytical assays based on colorimetric analysis. The results can be visually verified to the unaided eye or interpreted by a reader [81]. Nevertheless, fluorescence or electrochemical methods have become more widespread and attractive because of their high accuracy, sensitivity and lower limit of detection. Further, electrochemistry is less subject to the interference compounds exposed in the biological specimens, because it is not affected by ambient lighting conditions [21]. Colorimetric revealing has been expansively applied due to its simplicity and compatibility with cameras. Mobile phones are accessories widely available, allowing be coupled, and so, they are very suitable for incorporation into portable microfluidic devices. Their rapid improvement of hardware and software, high-resolution cameras, processing power and worldwide coverage of wireless internet network connection can facilitate diagnostic access, permit continuous monitoring of health parameters and promote increased surveillance notifications. This way, it is possible to do geo-timed reports and tracking of data automated providing governments with statistical information for clinical and epidemiological impact evaluation and counter-measures policies implementation. In fact, 3D printers and smartphones are instruments that are revolutionizing the future of lab-on-chip platform [82].

Other innovative actuation principle is centrifugal microfluidic that taking advantage of the forces acting on liquids in rotating chips. A spindle motor is necessary to press the fluid in the microfluidic chip. The centrifugal systems are particularly important for tasks involving separation of particles in suspension, as even small differences in density between solid parts and surrounding medium will result in sedimentation. It allows to perform the fluid manipulation within operational cartridges without the need of external micropumps and microvalves or previous sample preparation, as in the case to extract cell free plasma from whole blood [83]. Many challenges have been solved requiring only little of user interaction. These emergent microfluidic systems with integrated sensing, also termed lab-on-a-disc, are typically based on optical techniques, for example, absorbance, fluorescence or imaging. Optical readout with movable instrumentation are a successful detection and ensures several advantages: non-contact, high sensitivity, and the availability of optical components such as lasers and photo detectors or even constituents developed for optical disc drives [84].

2.5. Conclusions

Although great advances have been achieved in the development of PoCT immunosensors applied to health that facilitated the diagnosis of many diseases, control and handling more effectively, allowing analysis or multi-analysis more quickly, more remains to be done to make PoCT a practical devices for clinical routine. Carbon nanotubes, graphene metallic and magnetic nanoparticles nanostructures are examples of nanomaterials that have been widely used for electrochemical PoCTs, improving the amperometric transductions by promoting increase on the electron transfer and offer better electrocatalytic activity. Additionally, due to the large superficial area of nanomaterials, they are able result in increase on electroactive surface area, implying a high sensitivity for PoCTs. While many challenges still need to be overcome, the focus on PoCT immunosensor researches have grown exponentially on the last few decades. Many advantages make them ideal analytical methods: the phlebotomy step is avoided and replaced by a simpler and safer procedure; the collection of capillary blood with a few microliters can be performed on bedside; turnaround time of the diagnostic cycle is dramatically reduced and the results can be immediately informed to the patient, possibiliting the decentralization of outpatient services; and also the coupling with technologies for mobile phones and similar devices is possible.

Acknowledgments

Authors thank the National Council of Technological and Scientific Development (CNPq), Brazilian agency. Diego G. A. Cabral, Gilvânia M. de Santana, Paula A. B. Ferreira, Anne K. S. Silva, Erika K. G. Trindade, postgraduate students are grateful to FACEPE (Brazil) for their scholarships.

References

[1]. E. C. Rama, A. Costa-García, Screen-printed Electrochemical Immunosensors for the Detection of Cancer and Cardiovascular Biomarkers, *Electroanalysis*, Vol. 28, Issue 8, Aug. 2016, pp. 1700-1715.

[2]. C. Florkowski, A. Don-Wauchope, N. Gimenez, K. Rodriguez-Capote, J. Wils, A. Zemlin, Point-of-care testing (POCT) and evidence-based laboratory medicine (EBLM) – does it leverage any advantage in clinical decision making?, *Crit. Rev. Clin. Lab. Sci.*, Vol. 54, Issue 7-8, November 2017, pp. 471-494.

[3]. X. Fu, Z. Cheng, J. Yu, P. Choo, L. Chen, J. Choo, A SERS-based lateral flow assay biosensor for highly sensitive detection of HIV-1 DNA, *Biosens. Bioelectron.*, Vol. 78, 2016, pp. 530-537.

[4]. K. M. Koczula, A. Gallotta, Lateral flow assays, *Essays Biochem.*, Vol. 60, Issue 1, 2016, pp. 111-120.

[5]. M. Sajid, A. N. Kawde, M. Daud, Designs, formats and applications of lateral flow assay: A literature review, *J. Saudi Chem. Soc.*, Vol. 19, Issue 6, 2015, pp. 689-705.

[6]. E. B. Bahadır, M. K. Sezgintürk, Lateral flow assays: Principles, designs and labels, *TrAC – Trends Anal. Chem.*, Vol. 82, 2016, pp. 286-306.

[7]. D. Quesada-González, A. Merkoçi, Nanoparticle-based lateral flow biosensors, *Biosens. Bioelectron.*, Vol. 73, Nov. 2015, pp. 47-63.

[8]. J.-H. Lee, *et al.*, Multiplex diagnosis of viral infectious diseases (AIDS, hepatitis C, and hepatitis A) based on point of care lateral flow assay using engineered proteinticles, *Biosens. Bioelectron.*, Vol. 69, July 2015, pp. 213-225.

[9]. X. Gong, *et al.*, A review of fluorescent signal-based lateral flow immunochromatographic strips, *J. Mater. Chem. B*, Vol. 5, Issue 26, July 2017, pp. 5079-5091.

[10]. C. Kokkinos, A. Economou, M. I. Prodromidis, Electrochemical immunosensors: Critical survey of different architectures and transduction strategies, *Trends Anal. Chem.*, Vol. 79, May 2016, pp. 88-105.

[11]. N. S. Lipman, L. R. Jackson, L. J. Trudel, F. Weis-Garcia, Monoclonal versus polyclonal antibodies: Distinguishing characteristics, applications, and information resources, *ILAR J.*, Vol. 46, Issue 3, January 2005, pp. 258-268.

[12]. R. A. S. Couto, J. L. F. C. Lima, M. B. Quinaz, Recent developments, characteristics and potential applications of screen-printed electrodes in pharmaceutical and biological analysis, *Talanta*, Vol. 146, January 2016, pp. 801-814.

[13]. N. Gan, X. Du, Y. Cao, F. Hu, T. Li, Q. Jiang, An Ultrasensitive electrochemical immunosensor for HIV p24 based on $Fe_3O_4@SiO_2$ nanomagnetic probes and nanogold colloid-labeled enzyme-antibody copolymer as signal tag, *Materials (Basel).*, Vol. 6, Issue 4, March 2013, pp. 1255-1269.

[14]. L. Suresh, P. K. Brahman, K. R. Reddy, J. S. Bondili, Development of an electrochemical immunosensor based on gold nanoparticles incorporated chitosan biopolymer nanocomposite film for the detection of prostate cancer using PSA as biomarker, *Enzyme Microb. Technol.*, Vol. 112, May 2018, pp. 43-51.

[15]. M. M. P. S. Neves, M. B. González-García, H. P. A. Nouws, A. Costa-García, Celiac disease detection using a transglutaminase electrochemical immunosensor fabricated on nanohybrid screen-printed carbon electrodes, *Biosens. Bioelectron.*, Vol. 31, Issue 1, January 2012, pp. 95-100.

[16]. B. V. M. Silva, B. A. G. Rodríguez, G. F. Sales, M. P. T. Sotomayor, R. F. Dutra, An ultrasensitive human cardiac troponin T graphene screen-printed electrode based on electropolymerized-molecularly imprinted conducting polymer, *Biosens. Bioelectron.*, Vol. 77, 2016, pp. 978-985.

[17]. Y. Xu, S. Yang, W. Shi, Fabrication of an immunosensor for cardiac troponin I determination, *Int. J. Electrochem. Sci.*, Vol. 12, Issue 9, 2017, pp. 7931-7940.

[18]. A. Afkhami, P. Hashemi, H. Bagheri, J. Salimian, A. Ahmadi, T. Madrakian, Impedimetric immunosensor for the label-free and direct detection of botulinum neurotoxin serotype A using Au nanoparticles/graphene-chitosan composite, *Biosens. Bioelectron.*, Vol. 93, Jul. 2017, pp. 124-131.

[19]. E. A. de Vasconcelos, N. G. Peres, C. O. Pereira, V. L. da Silva, E. F. da Silva, R. F. Dutra, Potential of a simplified measurement scheme and device structure for a low cost label-free point-of-care capacitive biosensor, *Biosens. Bioelectron.*, Vol. 25, Issue 4, December 2009, pp. 870-876.

[20]. Z. Taleat, A. Khoshroo, M. Mazloum-Ardakani, Screen-printed electrodes for biosensing: A review (2008-2013), *Microchim. Acta*, Vol. 181, Issue 9-10, July 2014, pp. 865-891.

[21]. L. Syedmoradi, M. Daneshpour, M. Alvandipour, F. A. Gomez, H. Hajghassem, K. Omidfar, Point of care testing: The impact of nanotechnology, *Biosens. Bioelectron.*, Vol. 87, January 2017, pp. 373-387.

[22]. M. Vidotti, R. Torresi, S. I. Córdoba De Torresi, Eletrodos modificados por hidróxido de níquel: Um estudo de revisão sobre suas propriedades estruturais e eletroquímicas visando suas aplicações em eletrocatálise, Eletrocromismo E baterias secundárias, *Quim. Nov.*, Vol. 33, Issue 10, 2010, pp. 2176-2186.

[23]. Q. Huang, *et al.*, Nanotechnology-based strategies for early cancer diagnosis using circulating tumor cells as a liquid biopsy, *Nanotheranostics*, Vol. 2, Issue 1, 2018, pp. 21-41.

[24]. Z. Farka, T. Juřík, D. Kovář, L. Trnková, P. Skládal, Nanoparticle-based immunochemical biosensors and assays: Recent advances and challenges, *Chem. Rev.*, Vol. 117, Issue 15, August 2017, pp. 9973-10042.

[25]. N. J. Wittenberg, C. L. Haynes, Using nanoparticles to push the limits of detection, *Wiley Interdiscip. Rev. Nanomedicine Nanobiotechnology*, Vol. 1, Issue 2, March 2009, pp. 237-254.

[26]. K. S. Krishna, Y. Li, S. Li, C. S. S. R. Kumar, Lab-on-a-chip synthesis of inorganic nanomaterials and quantum dots for biomedical applications, *Adv. Drug Deliv. Rev.*, Vol. 65, Issue 11-12, November 2013, pp. 1470-1495.

[27]. Z. Wang, Z. Dai, Carbon nanomaterial-based electrochemical biosensors: An overview, *Nanoscale*, Vol. 7, Issue 15, 2015, pp. 6420-6431.

[28]. A. Ghasemi, *et al.*, Carbon nanotubes in microfluidic lab-on-a-chip technology: Current trends and future perspectives, *Microfluid. Nanofluidics*, Vol. 21, Issue 9, 2017, pp. 1-19.

[29]. C.-M. Tîlmaciu, M. C. Morris, Carbon nanotube biosensors, *Front. Chem.*, Vol. 3, 2015, pp. 1-21.

[30]. A. Dasgupta, L. P. Rajukumar, C. Rotella, Y. Lei, M. Terrones, Covalent three-dimensional networks of graphene and carbon nanotubes: Synthesis and environmental applications, *Nano Today*, Vol. 12, 2017, pp. 116-135.

[31]. Z. Zhu, An overview of carbon nanotubes and graphene for biosensing applications, *Nano-Micro Lett.*, Vol. 9, Issue 3, 2017, pp. 1-24.

[32]. J. Seo, *et al.*, Immunosensor employing stable, Solid 1-amino-2-naphthyl phosphate and ammonia-borane toward ultrasensitive and simple point-of-care testing, *ACS Sensors*, Vol. 2, Issue 8, 2017, pp. 1240-1246.

[33]. A. C. M. S. Dias, S. L. R. Gomes-Filho, M. M. S. Silva, R. F. Dutra, A sensor tip based on carbon nanotube-ink printed electrode for the dengue virus NS1 protein, *Biosens. Bioelectron.*, Vol. 44, Issue 1, 2013, pp. 216-221.

[34]. B. V. M. Silva, I. T. Cavalcanti, M. M. S. Silva, R. F. Dutra, A carbon nanotube screen-printed electrode for label-free detection of the human cardiac troponin T, *Talanta*, Vol. 117, 2013, pp. 431-437.

[35]. E. Sánchez-Tirado, C. Salvo, A. González-Cortés, P. Yáñez-Sedeño, F. Langa, J. M. Pingarrón, Electrochemical immunosensor for simultaneous determination of interleukin-1 beta and tumor necrosis factor alpha in serum and saliva using dual screen printed electrodes modified with functionalized double-walled carbon nanotubes, *Anal. Chim. Acta*, Vol. 959, 2017, pp. 66-73.

[36]. S. L. R. Gomes-Filho, A. C. M. S. Dias, M. M. S. Silva, B. V. M. Silva, R. F. Dutra, A carbon nanotube-based electrochemical immunosensor for cardiac troponin T, *Microchem. J.*, Vol. 109, 2013, pp. 10-15.

[37]. S. Ma, *et al.*, Interaction processes of ciprofloxacin with graphene oxide and reduced graphene oxide in the presence of montmorillonite in simulated gastrointestinal fluids, *Sci. Rep.*, Vol. 7, Issue 1, 2017, pp. 1-11.

[38]. R. Raccichini, A. Varzi, S. Passerini, B. Scrosati, The role of graphene for electrochemical energy storage, *Nat. Mater.*, Vol. 14, Issue 3, 2015, pp. 271-279.

[39]. H. Sadegh, Development of graphene oxide from graphite: A review on synthesis, characterization and its application in wastewater treatment, *Rev. Adv. Mater. Sci.*, Vol. 49, 2017, pp. 38-43.

[40]. L. G. Guex, *et al.*, Experimental review: Chemical reduction of graphene oxide (GO) to reduced graphene oxide (rGO) by aqueous chemistry, *Nanoscale*, Vol. 9, Issue 27, 2017, pp. 9562-9571.

[41]. J. Yukird, T. Wongtangprasert, R. Rangkupan, O. Chailapakul, T. Pisitkun, N. Rodthongkum, Label-free immunosensor based on graphene/polyaniline nanocomposite for neutrophil gelatinase-associated lipocalin detection, *Biosens. Bioelectron.*, Vol. 87, 2017, pp. 249-255.

[42]. P. K. Drain, *et al.*, Diagnostic point-of-care tests in resource-limited settings., *Lancet. Infect. Dis.*, Vol. 14, Issue 3, March 2014, pp. 239-249.

[43]. B. V. M. Silva, B. A. G. Rodríguez, G. F. Sales, M. D. P. T. Sotomayor, R. F. Dutra, An ultrasensitive human cardiac troponin T graphene screen-printed electrode based on electropolymerized-molecularly imprinted conducting polymer, *Biosens. Bioelectron.*, Vol. 77, 2016, pp. 978-985.

[44]. S. X. Lee, H. N. Lim, I. Ibrahim, A. Jamil, A. Pandikumar, N. M. Huang, Horseradish peroxidase-labeled silver/reduced graphene oxide thin film-modified screen-printed electrode for detection of carcinoembryonic antigen, *Biosens. Bioelectron.*, Vol. 89, 2017, pp. 673-680.

[45]. G. Doria, *et al.*, Noble metal nanoparticles for biosensing applications, *Sensors*, Vol. 12, Issue 12, Feb. 2012, pp. 1657-1687.

[46]. Y. Li, H. J. Schluesener, S. Xu, Gold nanoparticle-based biosensors, *Gold Bull.*, Vol. 43, Issue 1, March 2010, pp. 29-41.

[47]. H. Malekzad, P. Sahandi Zangabad, H. Mirshekari, M. Karimi, M. R. Hamblin, Noble metal nanoparticles in biosensors: recent studies and applications, *Nanotechnol. Rev.*, Vol. 6, Issue 3, January 2017, pp. 301-329.

[48]. M. Holzinger, A. Le Goff, S. Cosnier, Nanomaterials for biosensing applications: A review, *Front. Chem.*, Vol. 2, 2014, 63.

[49]. N. Li, P. Zhao, D. Astruc, Anisotropic gold nanoparticles: synthesis, properties, applications, and toxicity, *Angew. Chemie Int. Ed.*, Vol. 53, Issue 7, February 2014, pp. 1756-1789.

[50]. F. Inci, *et al.*, Multitarget, quantitative nanoplasmonic electrical field-enhanced resonating device (NE ^{2} RD) for diagnostics, *Proceedings of Natl. Acad. Sci.*, Vol. 112, Issue 32, August 2015, pp. E4354-E4363.

[51]. S. Kumar, W. Ahlawat, R. Kumar, N. Dilbaghi, Graphene, carbon nanotubes, zinc oxide and gold as elite nanomaterials for fabrication of biosensors for healthcare, *Biosens. Bioelectron.*, Vol. 70, August 2015, pp. 498-503.

[52]. M. Jacobs, A. Panneer Selvam, J. E. Craven, S. Prasad, Antibody-conjugated gold nanoparticle-based immunosensor for ultra-sensitive detection of troponin-T, *J. Lab. Autom.*, Vol. 19, Issue 6, December 2014, pp. 546-554.

[53]. S. Sabouri, H. Ghourchian, M. Shourian, M. Boutorabi, A gold nanoparticle-based immunosensor for the chemiluminescence detection of the hepatitis B surface antigen, *Anal. Methods*, Vol. 6, Issue 14, June 2014, pp. 5059-5066.

[54]. R. El-Dessouky, M. Georges, H. M. E. Azzazy, Silver nanostructures: Properties, synthesis, and biosensor applications, *ACS Symposium Series*, Vol. 1112, 2012, pp. 359-404.

[55]. X. Lu, M. Rycenga, S. E. Skrabalak, B. Wiley, Y. Xia, Chemical synthesis of novel plasmonic nanoparticles, *Annu. Rev. Phys. Chem.*, Vol. 60, 2009, pp. 167-192.

[56]. Z. Farka, T. Juřík, D. Kovář, L. Trnková, P. Skládal, Nanoparticle-based immunochemical biosensors and assays: Recent advances and challenges, *Chem. Rev.*, Vol. 117, Issue 15, August 2017, pp. 9973-10042.

[57]. N. Hao, *et al.*, An electrochemical immunosensing method based on silver nanoparticles, *J. Electroanal. Chem.*, Vol. 656, Issues 1-2, 2011, pp. 50-54.

[58]. S. Felici, *et al.*, Towards a model of electrochemical immunosensor using silver nanoparticles, *Procedia Technol.*, Vol. 27, January 2017, pp. 155-156.

[59]. X. Che, R. Yuan, Y. Chai, J. Li, Z. Song, J. Wang, Amperometric immunosensor for the determination of α-1-fetoprotein based on multiwalled carbon nanotube-silver nanoparticle composite, *J. Colloid Interface Sci.*, Vol. 345, Issue 2, May 2010, pp. 174-180.

[60]. Z. H. Ibupoto, N. Jamal, K. Khun, X. Liu, M. Willander, A potentiometric immunosensor based on silver nanoparticles decorated ZnO nanotubes, for the selective detection of d-dimer, *Sensors Actuators B Chem.*, Vol. 182, June 2013, pp. 104-111.

[61]. J. B. Haun, T.-J. Yoon, H. Lee, R. Weissleder, Magnetic nanoparticle biosensors, *Wiley Interdiscip. Rev. Nanomedicine Nanobiotechnology*, Vol. 2, Issue 3, May 2010, pp. 291-304.

[62]. J. B. Haun, T.-J. Yoon, H. Lee, R. Weissleder, Magnetic nanoparticle biosensors, *Wiley Interdiscip. Rev. Nanomedicine Nanobiotechnology*, Vol. 2, Issue 3, May 2010, pp. 291-304.

[63]. L. Syedmoradi, M. Daneshpour, M. Alvandipour, F. A. Gomez, H. Hajghassem, K. Omidfar, Point of care testing: The impact of nanotechnology, *Biosens. Bioelectron.*, Vol. 87, January 2017, pp. 373-387.

[64]. M. Martín, *et al.*, Rapid Legionella pneumophila determination based on a disposable core-shell Fe_3O_4 @poly(dopamine) magnetic nanoparticles immunoplatform, *Anal. Chim. Acta*, Vol. 887, August 2015, pp. 51-58.

[65]. V. Singh, S. Krishnan, Voltammetric Immunosensor Assembled on carbon-pyrenyl nanostructures for clinical diagnosis of type of diabetes, *Anal. Chem.*, Vol. 87, Issue 5, March 2015, pp. 2648-2654.

[66]. Y. Zhang, T. R. Nayak, H. Hong, W. Cai, Biomedical applications of zinc oxide nanomaterials, *Curr. Mol. Med.*, Vol. 13, Issue 10, December 2013, pp. 1633-1645.

[67]. S. Kumar, W. Ahlawat, R. Kumar, N. Dilbaghi, Graphene, carbon nanotubes, zinc oxide and gold as elite nanomaterials for fabrication of biosensors for healthcare, *Biosens. Bioelectron.*, Vol. 70, August 2015, pp. 498-503.

[68]. M. Marie, S. Mandal, O. Manasreh, An electrochemical glucose sensor based on zinc oxide nanorods, *Sensors*, Vol. 15, Issue 12, July 2015, pp. 18714-18723.

[69]. M. A. Seia, S. V. Pereira, M. A. Fernández-Baldo, I. E. De Vito, J. Raba, G. A. Messina, Zinc oxide nanoparticles based microfluidic immunosensor applied in congenital hypothyroidism screening, *Anal. Bioanal. Chem.*, Vol. 406, Issue 19, July 2014, pp. 4677-4684.

[70]. A. T. Giannitsis, Microfabrication of biomedical lab-on-chip devices. A review, *Est. J. Eng.*, Vol. 17, Issue 2, 2011, pp. 109-139.

[71]. Y. Temiz, R. D. Lovchik, G. V. Kaigala, E. Delamarche, Lab-on-a-chip devices: How to close and plug the lab?, *Microelectron. Eng.*, Vol. 132, January 2015, pp. 156-175.

[72]. M. Zarei, Advances in point-of-care technologies for molecular diagnostics, *Biosens. Bioelectron.*, Vol. 98, December 2017, pp. 494-506.

[73]. P. K. Drain, *et al.*, Diagnostic point-of-care tests in resource-limited settings., *Lancet. Infect. Dis.*, Vol. 14, Issue 3, March 2014, pp. 239-249.

[74]. F. A. Gomez, Paper microfluidics in bioanalysis, *Bioanalysis*, Vol. 6, Issue 21, November 2014, pp. 2911-2914.

[75]. R. Amin, *et al.*, 3D-printed microfluidic devices, *Biofabrication*, Vol. 8, Issue 2, June 2016, 22001.

[76]. J. Hu, *et al.*, Advances in paper-based point-of-care diagnostics, *Biosens. Bioelectron.*, Vol. 54, April 2014, pp. 585-597.

[77]. C. M. B. Ho, S. H. Ng, K. H. H. Li, Y.-J. Yoon, 3D printed microfluidics for biological applications., *Lab Chip*, Vol. 15, Issue 18, 2015, pp. 3627-3637.

[78]. K. Yamada, T. G. Henares, K. Suzuki, D. Citterio, Paper-based inkjet-printed microfluidic analytical devices, *Angew. Chemie Int. Ed.*, Vol. 54, Issue 18, April 2015, pp. 5294-5310.

[79]. M. Zarei, Advances in point-of-care technologies for molecular diagnostics, *Biosens. Bioelectron.*, Vol. 98, December 2017, pp. 494-506.

[80]. O. Strohmeier, *et al.*, Centrifugal microfluidic platforms: Advanced unit operations and applications, *Chem. Soc. Rev.*, Vol. 44, Issue 17, August 2015, pp. 6187-6229.

[81]. R. Burger, L. Amato, A. Boisen, Detection methods for centrifugal microfluidic platforms, *Biosens. Bioelectron.*, Vol. 76, February 2016, pp. 54-67.

Chapter 3

Sensors and Bioelectronics in the Kidney Replacement Therapy Applications

Marina V. Voinova, Leonid Y. Gorelik and Eugen I. Sokol

List of Abbreviations

aPPT - activated partial thrombo-plastin time

AQC - air quality control

BAW - bulk acoustic wave

BIA - bioimpedance analysis

BIS - bioimpedance spectroscopy

BSE - bovine serum albumin

CT - charge transfer

DS - dielectric spectroscopy

ECIS - electrical cell-substrate spectroscopy

ECW - extracellular water

EIT - electro-impedance tomography

EIS - electrical impedance spectroscopy

HD - hemodialysis

IBW - total body water

ICW - intracellular water

Marina V. Voinova

Chalmers University of Technology, Gothenburg, Sweden

IDWG - interdialytic weight gain

IF   - interstitial fluid

LOC   - lab-on-a-chip

MC   - microcystin

MF BIA - multifrequency bioimpedance analysis

MF BIS - multifrequency bioimpedance spectroscopy

MIP   - molecularly imprinted polymers

NPs   - nanoparticles

PA   - phase angle

PE   - polyethylene

PES   - polyethersulfone

PSF   - polysulfone

PT   - prothrombin time

QCM   - quartz crystal microbalance

QCM-D - quartz crystal microbalance with dissipation monitoring

QD   - quantum dot

RBC   - red blood cells

SAM   - self-assembled monolayers

SAW   - surface acoustic wave

SH-SAW - surface acoustic wave with horizontal polarization

SM   - supported membranes

TEG   - thromboelastography

UF   - ultrafiltration

3.1. Introduction

Kidney replacement therapy is undoubtedly a lifesaving method which should be ranked as one of the most prioritized of biomedical research in the future efforts [1-3]. The ultrafiltration (UF) process from blood flow to kidney is the central to the normal functioning and whole metabolism of living organism. In case of kidney failure, the blood filtration is achieved in hemodialysis (HD) apparatus where the optimal UF rate is a key parameter for the successful dialysis. It is crucial for the blood purification process, that the dialysis machines should be equipped with sensors and electronic devices controlling feedback parameters [4, 5].

Water, a universal solvent containing ions, proteins and nutrients, is a major component involved in the filtration of blood. The total fluid (in many cases called all body fluid or a total body water, TBW) in the human body is formed by extracellular water (ECW) and intracellular water (ICW). Plasma, a solution embedding blood cells, and the interstitial fluid (IF) filling up the small compartments between cells in tissues, are the most important ECW components. Monitoring of ICW and ECW in the process of controlled dialysis is a life-important procedure for renal patients. Being non-invasive method, the bioimpedance (BIA) [6-9] allows relatively simple and straightforward measurements of these patient's body parameters [6-21]. The development of the feedback systems, having a task of automatic control of the dialysis and ultrafiltration process to optimize the treatment parameters, requires sensors' measurements and computer modeling of the results [4, 5, 22].

To quantify the results of the measurements and to provide the correct interpretation of the experimental data, theory and adequate mathematical modeling should be developed and the numerical algorithm should be implemented in the control devices software. The current efforts in the theoretical modeling are represented and analyzed in both parts of the review.

The first part of the present review is dedicated to the multifrequency bioimpedance analysis (BIA), a non-invasive electrical method for monitoring blood and water volume of patients. A special attention is paid to the theoretical background of the BIA method with applications to the hemodialysis. The theoretical models and approaches based on Maxwell-Wagner interfacial polarization theory and Hanai-Koizumi theory of dielectric dispersions [23-28] (which are used for the design of software for bioelectronics supporting a multifrequency bioimpedance spectroscopy (MF BIS) for estimation of body composition, TBW, ECW and ICW) are reviewed as well as new experimental trends in impedance spectroscopy (EIS), in electrical cell substrate impedance sensing [31, 32] and theoretical developments.

In a healthy organism, blood purification is achieved in the natural filtering system in kidneys. When the natural filters are damaged or absent, the filtration occurs in dialysis machine via artificial membranes.

In hemodialysis, the blood cells brought in a close contact with synthetic, in most cases, polymer filters. Thus, a search of new coating for HD filters with improved biocharacteristics is an important challenge in HD-related studies. The search of new materials for the blood purification filters involves also question on how the components of blood interact with the filter membranes. Electrode surfaces modified with supported membranes, a combination of biological receptors and artificial polymeric coatings can serve as a convenient model system in there. Physical properties of the coating membrane and the interaction of blood cells with its surface can be deduced from the sensor response. Going in this direction, impedimetric and microgravimetric studies of living cells on artificial model membranes provide an important information on cell parameters and cell-surface interaction. In the review we pay attention to different aspects in electrical cell substrate impedance sensing (ECIS) and acoustical sensing with the quartz crystal microbalance (QCM). The combined EIS/ECIS and QCM methods in cell-surface monitoring are described as well.

Another common problem is blood clotting. Therefore, the monitoring of proteins' level (specifically, fibrin) in plasma is a necessary requirement for controlled HD. The information about dynamical changes due to the clot formation in viscoelastic properties of plasma could be obtained from measurements of viscoelastic properties of blood. Recently, a number of publications has reported QCM as an excellent sensor's method for the real time monitoring of blood plasma clotting and fibrin dynamics in solution.

In the third section of the current chapter we provide a short account for the experimental and theoretical developments of the QCM and QCM-D sensors used for the cells- artificial membranes interactions and in blood coagulation studies. The theoretical background together with basic models used for the QCM-D viscoelastic measurements of biological materials in fluids, is provided in the Section 3.3.6 of the chapter.

An important problem of water quality used for dialysis brought a new challenge for engineering of sensors to measure in real time an extremely low concentration of hazardous chemicals in solution. A short presentation of impedimetric (EIS/ECIS) and QCM-based sensors for water quality used for the dialysate preparation is included in the review, too (see Section 3.3.8).

3.2. Bioimpedance and Impedance Spectroscopy Methods in Biomedical Applications

3.2.1. Water Volumes Evaluation in Bioimpedance Technique

The control over a fluid removal and distribution in the patient's body is a primary task of dialysis management. The hydration of organism is directly related to the measure called "dry weight" [33, 39-48] and to the fluid changes in the extracellular and intracellular spaces [10, 11-21, 33-47]. The general purpose to control the dialysis process includes also data analysis, statistic and validation. Specifically, new achievements in this field include computations of intracellular, interstitial and intravascular volume changes during fluid management and correct evaluation of the 'dry weight' of the patients [10, 33, 39-47, 48]. In the review chapter [46], the authors give the following definition of the patient's dry weight: "Dry weight corresponds to the body weight of a person with normal extracellular fluid volume. In the context of hemodialysis, dry weight is the weight reached at the end of the dialysis session by patients who will remain free of orthostatic hypotension or hypertension until the next session. Clinicians are thus obliged to estimate the appropriate dry weight each individual patient should reach at the end of a dialysis session. If this weight is underestimated, the patients are at risk of various incidents ranging from simple yawning to death. Low dry weight also carries a permanent risk of hypotension, cramps, nausea, vomiting or ischemia. If this weight is overestimated, chronic hyperhydration can cause acute events including pulmonary edema, or hypertension, but also long-term consequences affecting cardiovascular morbidity and mortality. This important notion of dry weight is however quite problematic because it corresponds to a transient state, making it necessary to anticipate weight gain between two dialysis sessions and thus to reach a certain degree of dehydration at the end of each hemodialysis session."

Bioelectrical impedance analysis (BIA) is a convenient, budget and non-invasive experimental technique for the assessment of a fluid in the human organism [9], namely, a total body water (TBW) which includes the intracellular (ICW) and the extracellular (ECW) water, TBW = ICW + ECW [9-20]. The correct estimation of the hydration level related to ECW and ICW values is of key importance for the controlled hemodialysis time course when these parameters are time varying. The central point for the theory is the relationship between electrical measurable characteristics, which are the electrical resistance deduced from the amplitude of the complex electrical impedance $|\hat{Z}|$, and/or the phase angle, θ, and water volumes, ICW, ECW and TBW [13-15, 20, 35, 45, 46, 50]. These two parameters derived from BIA are typically used in healthcare research and a clinical practice [9]. The first parameter is mostly often taken for estimating body composition [9, 16-18, 20, 29, 30, 32, 37, 39, 50, 51, 55-59] and, in particular, water content relevant for the HD patients. However, this method has an essential drawback since it depends on specific equations related to the material parameters of the patients and represents an averaged result only. From the other hand, this well-developed experimental technique allows researchers and physicians non-expensive and safe method to collect and analyze a big amount of data obtained from the group of people, for example, according to races, gender or age. Another objective focuses on aid to the patients with specific diseases, (for example, diabetes mellitus and cardiovascular complications, often accompanying a kidney failure) sportsmen studies, dietary courses, and so on, with the aim to relate the electrical resistivity with changes in the physiological status of patients. Among others, there are two current challenges-first, the bioimpedance real-time monitoring of the patient status, second, how to make the test be individual for each patient [12-15, 21, 34, 35, 38, 39, 50, 55-59].

The phase angle (PA) measurement is used as an alternative parameter directly determined in BIA [9, 55]. This characteristic is independent of body height and weight and calculated as the arctangent of the experimentally measured ratio of the reactance to resistance [9]. Although the interpretation of the physiological meaning of the phase angle remains unclear, according to the number of current publications (see, for the review, [55]) it is the PA value which is considered as a perspective and powerful prognostic tool, in particular, for maintenance hemodialysis patients) [55].

The most important question is how to relate the electrical measurable characteristics, the electrical resistance deduced from the electrical impedance measurements, to the body fluid volumes, namely, the V_{ICW}, V_{ECW} and V_{TBW}. In the Section 3.2.3 we provide a brief account of the current results in this research.

The impedimetric measurements at different frequencies are generally classified as belonging to the dielectric spectroscopy (DS). Dielectric spectra are usually obtained by applying the electric field of varying frequency to biological objects of different nature. The electrical impedance spectroscopy (EIS) is a method applied to the biological system of living cells, suspensions or tissues, which enables to obtain characteristics of material measured in a certain frequency range [9, 31, 32, 60-63]. This method provides valuable information about dielectric properties of biological tissues [9, 29, 30]. Dielectric spectroscopy became a viable tool not only for tissue characterization but also for biosensors' analysis of cells physiology [31, 60-63]. Electrical impedance measurement of tissues [29, 30, 56, 59] and biological cells is widely accepted now as a label-free, non-invasive analytical method widely used to assess living cell functions [60-63]. The contemporary impedance technique includes also the electrical cell substrate impedance sensing (ECIS) (see, for the review, [31]).

The important statement should be made at this point. Despite of the diversity of above mentioned approaches, the common theoretical basis of them is based on the general Maxwell-Wagner theory of interfacial polarization and on its contemporary development, in particular, known as "mixture models". For example, a theoretical background behind the working principle of the BIA spectroscopy analysis (used, in particular, in Xitron-manufactured bioimpedance devices [64]) is based on Maxwell's theory of interfacial polarization and its further developments in models of dielectric dispersion by Hanai-Kouzumi. Simultaneously, the approach based on Maxwell-Wagner-Hanai "mixture models" [23-28, 66-68] is a theoretical platform for the analysis of electrical properties of cells in suspension or single cells in electrical cell substrate sensing (ECIS) spectral analysis [31, 60-63]. Because of the importance of this approach both in BIA and EIS/ECIS, we provide a step-by-step consideration of above mentioned theoretical models and show how the results of calculated electrical characteristics can be related to the assessment of body fluid.

3.2.2. On the Theory of Interfacial Polarization

In the analysis of disperse system with stratified structure, Hanai [27] derived the expression for the complex electric conductivity of the system as a function of volume fraction of the disperse phase Φ.

Let us consider a heterogeneous system consisting of dielectric dispersion (phase I, index "p", embedded into phase II, or medium, index "m"). The complex permittivity is

$$\varepsilon^*(\omega) = \varepsilon' - i\varepsilon'' \tag{3.1}$$

and, thus, for each phase:

$$\varepsilon_m^* = \varepsilon_m' - i\varepsilon_m'', \tag{3.2}$$

$$\varepsilon_p^* = \varepsilon_p' - i\varepsilon_p''$$

Here, ε' denotes real part of the permittivity, ε'' corresponds to the imaginary part of this value. By introducing a volume fraction ϕ of the dispersed phase (index "p"), one can obtain for the complex permittivity ε^* of mixture:

$$\varepsilon^* = \varepsilon_m^* \frac{\varepsilon_p^*}{\varepsilon_p^* + \phi(\varepsilon_m^* - \varepsilon_p^*)} \tag{3.3}$$

and for the complex conductivity σ^* of mixture:

$$\sigma^* = \sigma_m^* \frac{\sigma_p^*}{\sigma_p^* + \phi(\sigma_m^* - \sigma_p^*)} \tag{3.4}$$

For the limit case of low frequency ($\omega \to 0$) (index "L") and high frequency ($\omega \to \infty$) (index "H"), from (3.3) and (3.4) one can get for the permittivity

$$\varepsilon_H = \varepsilon_m \frac{\varepsilon_p}{\varepsilon_p + \phi(\varepsilon_m - \varepsilon_p)} \tag{3.5}$$

$$\varepsilon_L = \frac{\varepsilon_m \sigma_p^2 + \phi(\varepsilon_p \sigma_m^2 - \varepsilon_m \sigma_p^2)}{[\sigma_p + \phi(\sigma_m - \sigma_p)]^2} \tag{3.6}$$

and for the conductivity, respectively:

$$\sigma_H = \frac{\sigma_m \varepsilon_p^2 + \phi(\sigma_p \varepsilon_m^2 - \sigma_m \varepsilon_p^2)}{[\varepsilon_p + \phi(\varepsilon_m - \varepsilon_p)]^2} \tag{3.7}$$

$$\sigma_L = \sigma_m \frac{\sigma_p}{\sigma_p + \phi(\sigma_m - \sigma_p)} \tag{3.8}$$

For the system of n spherical particles each characterized with the resistivity k_p and placed in a medium of resistivity k_m, one can find Maxwell's expression (Maxwell 1881, p. 365) [23] for the effective resistivity. Considering the 'diluted'' dispersion case, i.e., when the spheres are far enough from each other and thus, may be treated as non-interacting particles, one can found for the resistivity of mixture, K:

$$K = \frac{2k_p + k_m + \phi(k_p - k_m)}{2k_p + k_m - 2(k_p - k_2)} k_m \qquad (3.9)$$

For the particular case, when $k_m \gg k_p$ from (3.9) one can get:

$$K \approx \frac{1-\phi}{1+2\phi} k_m \qquad (3.10)$$

and for $k_m \ll k_p$

$$K \approx \frac{2+\phi}{2(1-\phi)} k_m \qquad (3.11)$$

For the conductivity, respectively:

$$\sigma \approx \frac{(1+2\phi)}{1-\phi} \sigma_m, k_m \gg k_p \quad (3.12)$$

$$\text{and } \sigma \approx \frac{2(1-\phi)}{2+\phi} \sigma_m, k_m \ll k_p \qquad (3.13)$$

Among later important developments after Maxwell-Wagner's model, one should refer to Bruggeman theory [25], De la Rue and Tobias work [26], Fricke's model [67] for the non-spherical (ellipsoidal) particles, Looyenga's heterogeneous mixture theory [67], Hanai-Koizumi models [27, 28] and Asami- Hanai-Koizumi work [60, 68] as the most often used 'mixture models''.

The Wagner-Maxwell model does not hold for the concentrated dispersions. The generalized treatment of the problem has been developed by Bruggeman [25], Hanai [27], Hanai and Koizumi [28], also Asami, Hanai and Koizumi [60, 68].

Hanai-Koizumi theory of mixture

Rigorous calculation of complex permittivity and complex conductivity of water/oil mixtures in general case has been reported in [27]. The calculations are cumbersome, and readers can find the intermediate steps in the original paper [27]. We provide here the solution from Hanai's work [27] where the permittivity and conductivity of mixture are related as following in the system of two equations:

$$\frac{\{(\varepsilon\prime-\varepsilon_p')^2 + (\varepsilon''-\varepsilon_p'')^2\}\cdot\{\varepsilon_m'^2 + \varepsilon_m''^2\}^{1/3}}{\{(\varepsilon_m'-\varepsilon_p')^2 + (\varepsilon_m''-\varepsilon_p'')^2\}\cdot\{\varepsilon'^2 + \varepsilon''^2\}^{1/3}} = (1 - \phi)^2 \quad (3.14)$$

$$3\tan^{-1}\frac{(\varepsilon'-\varepsilon_p')(\varepsilon_m''-\varepsilon_p'')-(\varepsilon_m'-\varepsilon_p')(\varepsilon''-\varepsilon_p'')}{(\varepsilon'-\varepsilon_p')(\varepsilon_m'-\varepsilon_p') + (\varepsilon''-\varepsilon_p'')(\varepsilon_m''-\varepsilon_p'')} = \tan^{-1}\frac{\varepsilon'\varepsilon_m''-\varepsilon_m'\varepsilon''}{\varepsilon'\varepsilon_m' + \varepsilon_m''\varepsilon''} \quad (3.15)$$

The system of equations (3.14, 3.15) has been further analyzed for various conditions, i.e., for the special relations between conductivities of particles/medium as well as for the high and low frequencies limits.

In particular, for the low frequency case ($\omega \to 0$) and for the conditions $\varepsilon_m'' \gg \varepsilon_m'$, and $\varepsilon_p'' \gg \varepsilon_p'$ together with $\varepsilon'' \gg \varepsilon'$, the equations (3.14), (3.15) are reduced to the simpler expressions [27]. The low frequency limit, σ_L, for the conductivity of mixture can be found from following equation [27]:

$$\frac{\sigma_L-\sigma_p}{\sigma_m}\left(\frac{\sigma_m}{\sigma_L}\right)^{1/3} \approx 1 - \phi \quad (3.16)$$

For a special case, when $\sigma_L \gg \sigma_p$, formula (3.16) is reduced to the equation:

$$\sigma_L \approx \sigma_m(1 - \phi)^{3/2} \quad (3.17)$$

This equation has been extensively used in colloidal and membrane sciences as one of the 'mixture models'' as well as for the calculations of electrical characteristics of biological tissues [11-15].

In the next section we consider the applications of this equation in the theory of bioelectrical impedance and water content evaluation.

3.2.3. Multifrequency Bioelectrical Impedance /Spectroscopy (MF-BIA/BIS). Bioimpedance Analysis in Water Content Measurements

The important next step is to understand how to relate the results of the 'mixture model' theory with macroscopic electrical measurable characteristics, i.e. the electrical resistance deduced from the electrical impedance amplitude and, then, to the body fluid volumes, namely, the V_{ICW}, V_{ECW} and V_{TBW}. In the most publications on bioimpedance body fluid measurements [11-15], the following relation between the resistivity of mixture of non-conductive and conductive elements and volume characteristics is used:

$$\rho = \frac{\rho_0}{(1-\upsilon)^{3/2}} \qquad (3.18)$$

In this expression, ρ_0 is the actual resistivity of a conductive material, υ is a volumetric concentration of the nonconductive material in the mixture:

$$\upsilon_{LF} = 1 - \frac{V_{ECW}}{V_{Total}} \qquad (3.20)$$

and

$$\upsilon_{HF} = 1 - \frac{V_{ECW} + V_{ICW}}{V_{Total}}, \qquad (3.21)$$

where V_{Total} is the total volume.

In BIS measurements of body fluid volumes, this formula is used in combination with the body material parameters. For example, in papers [13, 34], it was shown that

$$V_{ECW\,Xitron} = \left\{ \frac{\rho_{ECW} \cdot K_B \cdot H^2 \sqrt{W_t}}{\sqrt{D_b} \cdot R_0} \right\}^{2/3}, \qquad (3.22)$$

where K_B is so-called "shape-factor" introduced for the correction of measurements between wrist and ankle relating the geometrical proportions of the leg, arm, trunk and height.

For the assessment of the extracellular water volume, Matthie [13, 14] applied the theory of Hanai [27] for the mixture of two conductive fluids with different conductivities, σ_{ECW} and σ_{ICW}, respectively,

$$(1 - \upsilon_{ICW}) = \frac{\sigma_{TBW} - \sigma_{ICW}}{\sigma_{ECW} - \sigma_{ICW}} \cdot \left\{ \frac{\sigma_{ECW}}{\sigma_{TBW}} \right\}^{1/3}, \qquad (3.23)$$

which yields:

$$V_{ICW\,Xitron} = V_{ECW\,Xitron} \left\{ \left[\frac{\rho_{TBW} \cdot R_0}{\rho_{ECW} \cdot R_\infty} \right]^{2/3} - 1 \right\} \qquad (3.24)$$

Here the total body resistivity ρ_{TBW} is given by the following relation:

$$\rho_{TBW} = \rho_{ICW} - (\rho_{ICW} - \rho_{ECW})\left(\frac{R_\infty}{R_0}\right)^{2/3}, \qquad (3.25)$$

where ρ_{ICW} is the intracellular resistivity [13, 14].

The equations (3.20-3.25) have been used by the author [13] to calculate the V_{ICW} value.

Both theories [11-15] which are widely used in bioimpedance practice, suggest a combination of the electrical circuit model resistances and the computation of the V_{ECW} volume. The above mentioned theoretical model has been used as a theoretical background for the software of bioimpedance devices manufactured by Xitron [64].

In the most cited publications [11-15, 32], the volume equations for ICW and ECW are related with the body material parameters, such as the body height (H), the body weight (W_t), the body density (D_b) and the electrical characteristics, which are the resistance $(R_0$ and $R_\infty)$ of the body fluid and the intracellular resistivity (ρ_{ICW}), extracellular resistivity (ρ_{ECW}) and the total body resistivity (ρ_{TBW}), respectively. The indices "0" and "∞" denote the low frequency (LF) and high frequency (HF) limits respectively which are usually obtained in the routine procedure from the extrapolation of the measured electrical impedance by using the Cole-Cole equation [65].

Measurements of multiple frequencies form a basis not only for the impedance spectroscopy [69] but also of new promising method of electroimpedance tomography (EIT) [70] where the choice of optimal frequencies is the important practical goal.

The ratio ECW/TBW measured with BIA devices allows to estimate the (approximate) amount of fluid for dialysis.

However, the bottle-neck is that electrical impedance method is unable to discriminate between the amount of interstitial fluid in the interdialytic weight gain (IDWG) and plasma volume [8, 10, 46]. Therefore, a further improvement in the theory is needed [71]. The opinion on how to distinguish the whole-body excess fluid from the hydration of particular tissues has been reported in [10, 72].

Another help is a usage of additional experimental methods to complete the body fluid analysis [38, 40]. For instance, in clinical practice, the Crit-Line monitor is used for continuous and non-invasive measurements of patient's hematocrit and calculation of changes in patient's intravascular blood volume [39].

Current publications on hematocrit in line with blood cells evaluation based on BIA are reported in [73, 74]. The example of new trends in HD research and clinical dialysis practice is BCM [75] which is another type of electronic devices used for the whole-body bioimpedance spectroscopy in dialysis.

3.2.4. Challenges in the Theory of Multifrequency Electrical Impedance Analysis

The theoretical basis for the electrical impedance measurements in biological tissues is still in progress. The dielectric properties of cells are strongly dependent on physiological conditions and can be affected by pathological state or damage of organs. The strike example provides the acute ischemia, or other heart diseases and diabetes mellitus, which are common complications of kidney failure. The difference in the dielectric properties between healthy and pathological tissues can be calculated as a function of frequency [29, 30]. The practical need comes also from newly developed multifrequency BIS [30, 59] techniques such as electrical impedance tomography (EIT) [70] and ECIS [60-63].

Recent trends include dielectric models of cellular structures in a wide frequency spectrum. In there, the dielectric properties of biological tissues have been modeled in the radio frequency and microwave diapason. Dielectric properties both of the electrically interacting and non-interacting biological cells have been calculated [76]. The EIT tomography experimental data (for various type of cells, including myocardial tissue, brain tissue among others, measured in the microwave spectrum from 0.2 to 6.0 GHz as well as in the low frequency region) have been compared with the models for the interpretation of measurements. The reasonable agreement was found at frequencies of 1-100 MHz. In the theoretical work [76] it was shown, that the calculated difference in dielectric properties between electrically non-interacting versus interacting cells is inversely dependent on frequency. The appearance of new advanced mathematical models for electrically interacting cells is a new and important development [76-78] comparing with a classical theory of interfacial polarization and Hanai model of dielectric dispersion where cells were considered as non-interactive objects.

3.3. The Acoustic and Impedimetric Methods in Biomedical Applications

Acoustic measurements of the surface mass adsorbed onto the surface of quartz crystal microbalance (QCM) in the air (or vacuum) is based on the linear dependence of the changes in resonance frequency of the QCM Δf caused by a thin solid film of mass m [79]:

$$\Delta f = -f_0 \frac{m}{m_q} \qquad (3.26)$$

This linear (Sauebrey) relation lies at a basis of microgravimetric principle of QCM, operated in vacuum or gases. In gases, QCM has been intensively used for the detection of volatile compounds in the air quality control (AQC) (see the selected examples in [83-85]). Since the detected mass is "anonymous", the surface of the resonator must be functionalized with a specific analytes for the selectivity of chemical binding. In a similar way, when the QCM operates in liquids [80-82], the adsorption of molecules from the solution requires the surface pre-treatment of the QCM for the specific target binding.

During last two decades, a conventional quartz crystal microbalance, originally used for gravimetrical measurements of thin solid films in gases, evolved into a biosensor's analytical tool operating also in liquids. The important achievement in the field was finding that viscous damping measurement may bring additional information about properties of tested liquid [80-82, 86-88]. (The literature on QCM is vast so for the further reading we can recommend, for example, the surveys published in [89-91]).

It was found [80], that in case of Newtonian bulk liquids, the resonance frequency shift of the QCM depends on the product of density ρ and shear viscosity η of tested liquid:

$$\Delta f = -\frac{1}{2\pi m_q} \sqrt{\frac{\rho \eta \omega}{2}} \qquad (3.27)$$

Two crucial steps were i) a construction of a QCM-D (a quartz crystal microbalance with dissipation ΔD monitored in parallel with Δf) [88, 92-95] and ii) advances in surface chemistry allowing a functionalization of the quartz crystal resonator with polymers, supported membranes or biomolecular receptors sensing the analytes of various nature (see, for the review, in the recently published book [91], also in the publications in [96]). Currently, QCM and QCM -D devices are popular acoustic sensors used, in particular, for environmental, biological and biomedical purposes [96]. Practical applications of QCM-based sensors include also the detection of traces of heavy metals and other hazardous contaminants in water (see also, in Section 3.3.8).

Basic scheme and the electric characteristics of the quartz crystal microbalance are shown on Fig. 3.1.

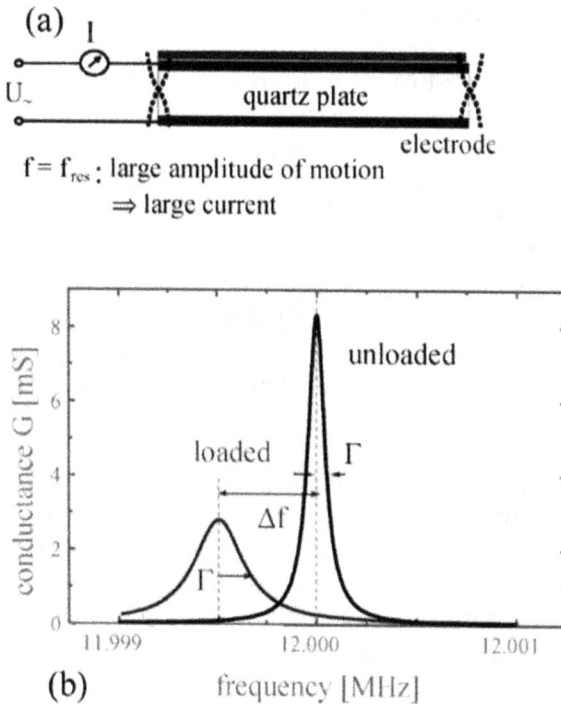

Fig. 3.1. (a) A sketch of the quartz crystal microbalance, (b) electrical characteristics of loaded (with adsorbed mass) and unloaded QCM (reproduced from [97], with permission).

Characteristics of biological materials measured with the QCM demonstrate a deviation from linear (Sauerbrey) behaviour. The reason of this non-Sauerbrey behaviour is a viscoelasticity (softness) of biomaterials (see, for review, for example, [91, 96-113]) and the viscoelastic coupling between layers [114, 115].

Since late 1990s, the QCM has been suggested as a tool for measuring besides the surface mass also a viscoelasticity of polymer films and biomolecular layers [91, 95-102, 105-114]).

The soft and biological matter studies put new questions for the theoreticians, specifically, how to quantify the results of measurements and how to deduce the softness of the adsorbed material? The liquid phase measurements of soft biomaterials are typical in the applications of QCM as an analytical tool [89, 91, 95]. This brought also an important question of the correct physical interpretation of the measured QCM characteristics and necessity of mathematical modelling of the experimental results [91, 95-114].

Two steps of the pre-treatment are a coating of QCM surface with polymer or lipid layer and tethering of the receptive units which unable biosensor's and electrochemical biosensor's [116] applications of the QCM. Due to the success in surface functionalization of the QCM with biological molecules such as proteins and DNA, the quartz crystal microbalance became a useful analytical tool in biomedical research which allows to monitor the target components adsorbed from liquid in real time (since the list of relevant publications is very long, we refer here and further in the text to the database of [96]). The mode of signal transduction could be different depending on biological receptor specificity.

In this section, a short account for biomedical hemodialysis-related QCM applications is given. Most subsections are reviewing the experimental activity. Section 3.3.7 provides a theoretical background for the QCM studies of polymers and soft biological materials in liquids. A special attention is paid to the application of QCM and a combination of EIS/ECIS methods with the QCM for the control of water quality used for dialysis (Section 3.3.8).

3.3.1. Pre-Treatment of the QCM Surface in Biosensors' Applications

3.3.1.1. Supported Membranes and Biomimetic Strategies

Supported membranes (SM) technique received a considerable attention as a convenient method of biofunctionalization of inorganic substrates [117-119, 122-124] and, in particular, of quartz crystal microbalance [91, 96, 120, 121, 125-130]. Deposited onto the solid surface, SMs allow to build up artificial membranes of polymer- or lipid multilayers chemically attached to the solid state transducers [117-119]. Langmuir-Blodgett films, self-assembled monolayers (SAM), tethered membranes and lipid bilayers are among the most popular SMs techniques [117-119, 125- 130].

Different schemes of solid surfaces modification are shown on Fig. 3.2.

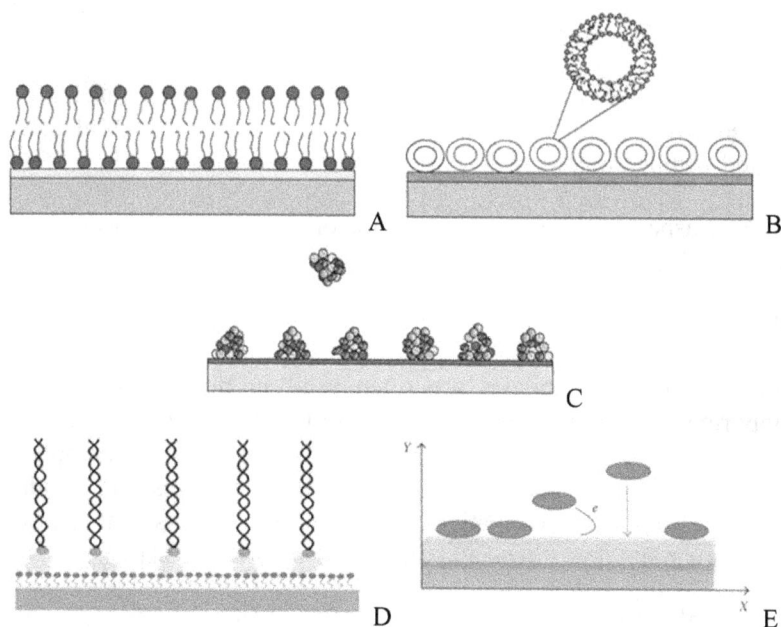

Fig. 3.2. A sketch of different variants of surface modification of the quartz crystal for biosensors' purposes; A-planar lipid bilayer membrane; B-surface-adhered lipid vesicles; C-protein adsorbed layer; D-DNA-based artificial membrane architecture; E-redox enzyme coating (reproduced from [147], with permission).

Lipid bilayer membranes deposited onto the solid surface of transducer, form a natural environment for biological molecules such as proteins, enzymes, antibodies and antigens [117-119]. These artificial membranes can mimic molecular architecture and functions of cell membranes. Also, since lipid bilayers are tolerant to biological membranes, it is possible to study with this technique the living cells-artificial membranes interaction [100, 117-119, 125-130]. There are exponentially growing number of reports on protein adsorption on supported lipid bilayer monitored with the QCM. The selected publications one can find, for example, in [131-139], in the specialized QCM reviews [90, 138] or in the publications of database [95].

The interaction of vesicles with different surfaces has both fundamental and applied aspects. Fundamental aspects include studies of physical and chemical forces involved in vesicles adhesion [144-146]. One of the important application is the spontaneous formation of the planar lipid bilayers on surfaces caused either by lipid vesicles rupture on various surfaces [117, 118, 120, 121, 124-126, 130, 134] or spreading of lipid layers after hydration [117, 118, 122].

Transformation of lipid vesicles to a bilayer on a gold substrate has been studied with QCM-D in [141]. In there, the experimental QCM-data were treated within Voigt element-based viscoelastic model (see Fig. 3.3).

Various aspects of liposomal behavior on surfaces have been studied in [140-146].

Biomimetics

The supported membranes are often used in biomimetics. Besides supported membranes, one should mention molecularly imprinted and bioimprinted polymers (MIP) [148-150], self-assembled monolayers (SAM), surface immobilized peptides, DNA enzymes and dendritic polymers used in combination with natural receptors, nanoparticles (NPs), quantum dots (QDs) and other nanomaterials [151].

Non-covalent molecular imprinting of polymers (MIP) is a new technique aimed to design artificial receptors or artificial antibodies that mimic related biological function of membrane components [150, 151]. In some biosensors, the imprinted polymers are used together with the natural receptors. The attachment of these units to SMs or SAMs (see, for example, in [150]) enables a specific recognition function and selectivity to the QCM. These features are often employed for drugs and toxins recognition.

Bioimprinting is a version of MIP technique where the proteins such as albumin [152] or enzymes (for instance, trypsin [150]) are printed from solution to the pre-coated surface of the device. Fig. 3.4 represents a sketch of two variants of QCM surface imprinting-an aqueous imprinting from solution and a crystal one. In there, enzymes are used as the model templates [150].

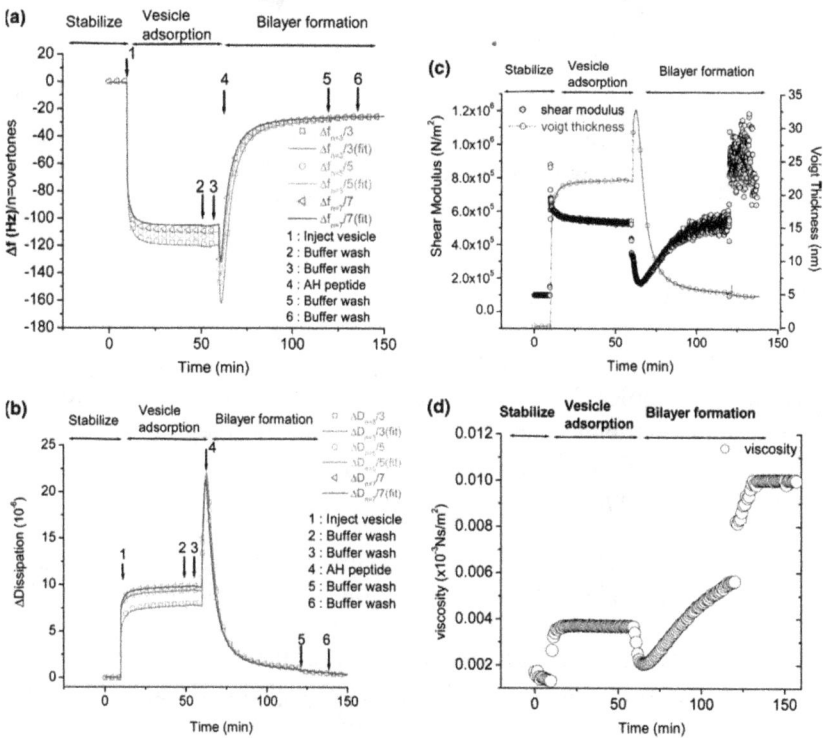

Fig. 3.3. Kinetics of lipid vesicles adsorption measured with QCM-D sensor. The experimental data (the shift in the resonance frequency (a) and in the dissipation (b)) were treated with the help of Voigt element-based viscoelastic model. (c) Changes in the effective shear modulus (open hexagon) and Voigt-based thickness as a function of time obtained from viscoelastic Voigt-Voinova model. (d) Shear viscosity as a function of time from the viscoelastic Voigt-Voinova model (reproduced from [141], with permission).

83

Basic principle of imprinting is shown on Fig. 3.4.

(a)   (b)

Aqueous Imprinting   Crystal Imprinting

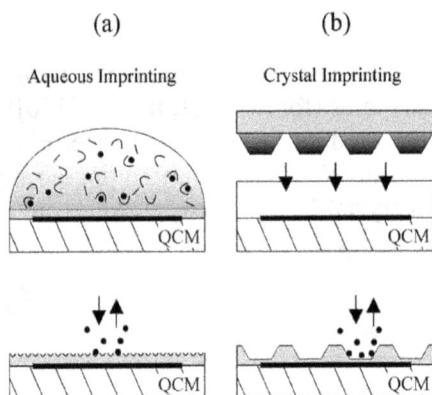

Fig. 3.4. Schematic illustration of two variants of the imprinting on the surface of pre-coated QCM: a-aqueous imprinting of proteins; b-crystal imprinting. (From [150], with permission).

DNA enzymes represent a new class of biomimetic materials with artificial receptors (DNAzymes). In this technique, enzymes are used in combination with DNA. This technique permits the surface coating with hybrid macromolecules enabling a specific recognition on molecular level (see, in particular, the example of biosensors based on DNAzymes designed for the detection of water toxins, in [153], also Section 3.3.8 of the current review).

DNA immobilization on chemically treated surface of QCM-D has been recently studied in [154], as well as in combination with vesicles [155].

The full review of the existing technologies for the SMs and biomimetics for biosensors is beyond the limit of current chapter and can be found by interested readers elsewhere (see, for example, in [117-119, 148-151]).

3.3.1.2. Lipid Vesicles Adhesion: Modelling Cell- Substrate Interaction

Adsorption of lipid vesicles onto the surface of QCM allows researchers to investigate specific interactions of cells with artificial surfaces [125, 156]. QCM-based studies of lipid vesicles have been reported the target-binding processes [142], the viscoelastic behavior upon transformation of vesicles to bilayer [141], the transmembrane helix dimerization [140] and the formation of phospholipid bilayer of nanostructured surfaces [130, 143]. Recent QCM-D study of complex membranes containing special peptides (AH peptides) has been published elsewhere [141] (see Fig. 3.5).

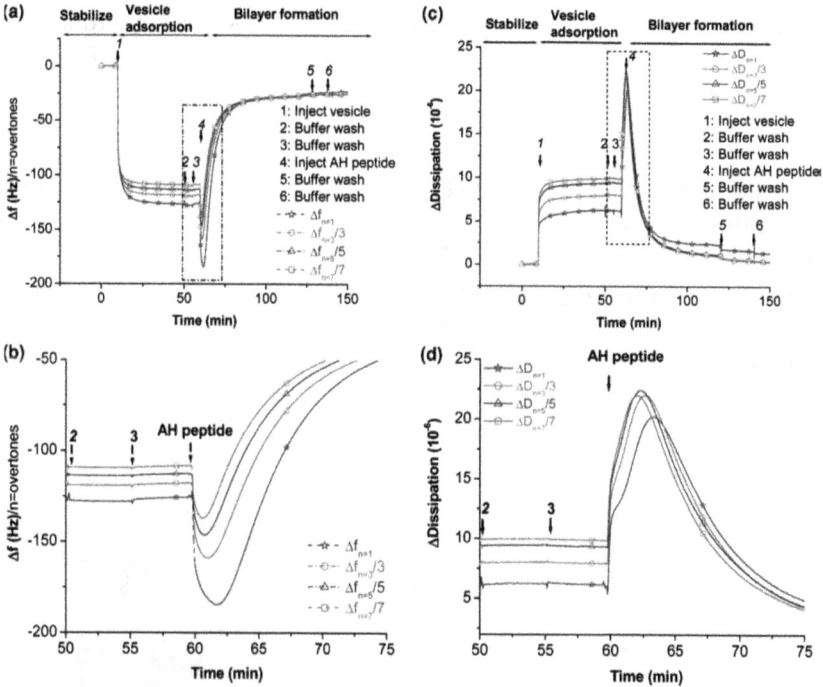

Fig. 3.5. Complex supported membranes studied with QCM-D. Special peptides (AH peptides) were added to study lipid vesicles stability, rupture and formation of lipid bilayer. The kinetics of structural transformation of adsorbed lipid vesicles was studied on different overtones versus time (reproduced from [141], with permission).

3.3.2. The Cells Adhesion to the Surfaces Measured with the QCM

The adhesion of living cells can be qualitatively studied with the QCM-D method [100, 104, 125, 128, 129, 133, 156-162] as well as in LOCs and microfluidic devices integrated with the QCM (see more in the Section 3.3.8). The adhesion of blood cells to the artificial surfaces is unwanted process in HD filtration. However, for the other type of cells, adhesion could be a very helpful factor. For example, a strong attachment of cells to the substrate is a requirement in cells culture and tissue engineering as well as in living cells -based sensors. The promising direction is measurement of living cells response to various stimuli such as addition of drugs or other chemical compounds. In this method, cells cultured directly on the quartz crystal (in QCM-D) or electrode surface (in ECIS) work as living sensors reacting on changes in the environment. These cells-based sensors are widely used now, for example, in drugs research and for water toxins detection (see more, in the Section 3.3.8).

The study of cells adhesion and spreading on artificial surfaces can provide valuable information on biocompatibility of the surface coatings. Recently, the real time QCM measurements of living cells have been published elsewhere [156-158]. The example of the QCM study of cells adhesion and spreading is shown on Fig. 3.6. In this work, a cytoskeleton modulator was added to disrupt the actin filaments and induce a reversible change in cell morphology.

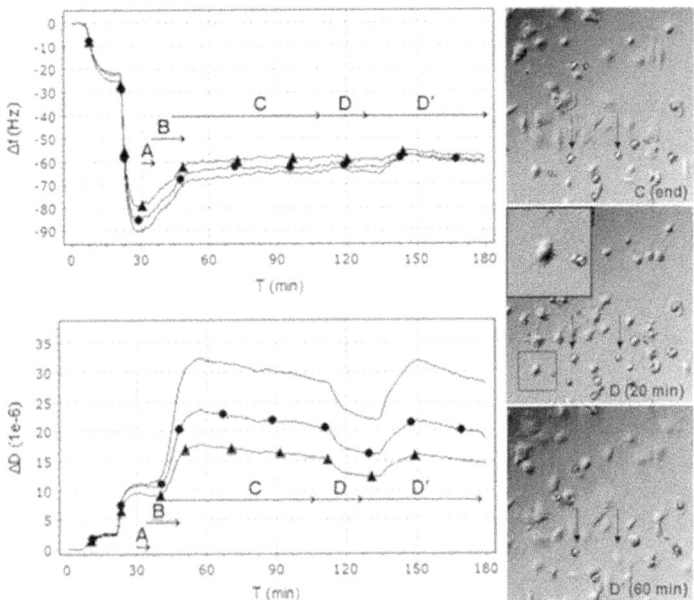

Fig. 3.6. To the left: the QCM-D real time measurements of living cells adhesion and spreading. To the right: parallel microscopy studies (from [158], with permission) showing the cells transformation with time.

One should point out the existing difficulties in the quantitative interpretation of the QCM results with cells. Since living cells layer is a multiparametric dynamic system, one could expect the time variation in cells mechanical (e.g., surface density) and viscoelastic (or rheological) properties as well as the morphological changes in cells response. Here the usage of different experimental methods (such as SPR, AFM and laser confocal microscopy) in parallel with the QCM-D measurements supported by adequate theory could be advantageous.

Although the rigorous theory has been developed for the QCM-D (see more, in the Section 3.3.6), it works very well for the relatively simple soft and homogeneous systems such as polymer layered films, proteins, lipid bilayer membranes and polyelectrolyte multilayers in liquids. For cells-based sensors, more theoretical work on modelling of biological cells dynamics in real time is needed.

Simultaneously, shape transformation of blood cells induced by flow in artificial microvascular networks [163] or by contact with surfaces [164] may provide valuable information on blood rheology in small capillaries [163] and cell-surface interaction [165, 166]. For example, red blood cells (RBC) behave as living-cell-sensors since easily change their shape on contact surfaces [164, 167]. It was proved [165-167] that the primary factor in RBC-surface interaction is the chemical composition of the artificial surfaces.

In particular, it was shown that for polyurethane and polysulfone membranes, the morphology of the adherent RBCs was different from shapes observed on poly-L-lysine coated supports. (The shape of the RBCs on the poly-L-lysine-coated substrate is a spherical cap, which can be explained as a consequence of strong adhesion forces between the membrane and the underlying polymer film).

The shapes of RBCs on polyurethane fibres is shown on Fig. 3.7 (experimentally observed RBC shapes, laser confocal microscopy) and Fig. 3.8 (computed RBC shapes), respectively.

The influence of morphological changes of RBCs to the dielectric spectroscopy (DS) characteristics has been confirmed in [32]. In there, a systematic study of DS sensitivity to cells morphology has been performed for different shapes of RBCs [32]. The study revealed that the characteristic DS frequency and the broadening parameter of the dielectric relaxation of interfacial polarization were highly specific in response to the shape of cells [32].

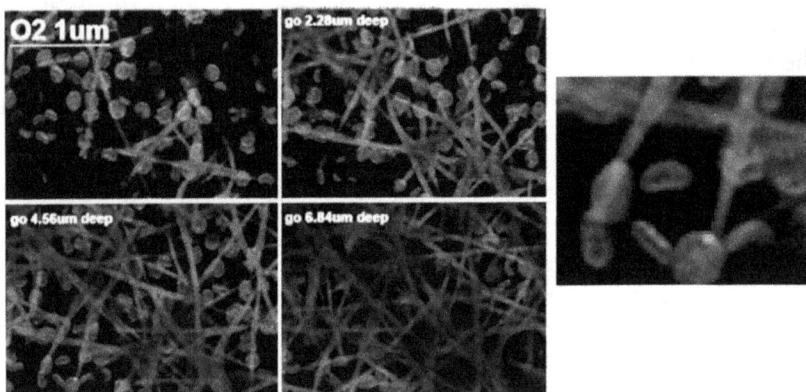

Fig. 3.7. *Left:* Confocal microscopy images of RBCs on oxygen plasma-treated polyurethane fibrous membrane taken at different depths from the polymer mesh surface (Zeiss LSCM, Meta 510, Zeiss). *Right:* A RBC bent over a fibre (a zoomed fragment of the image on the left) (from [167], with permission).

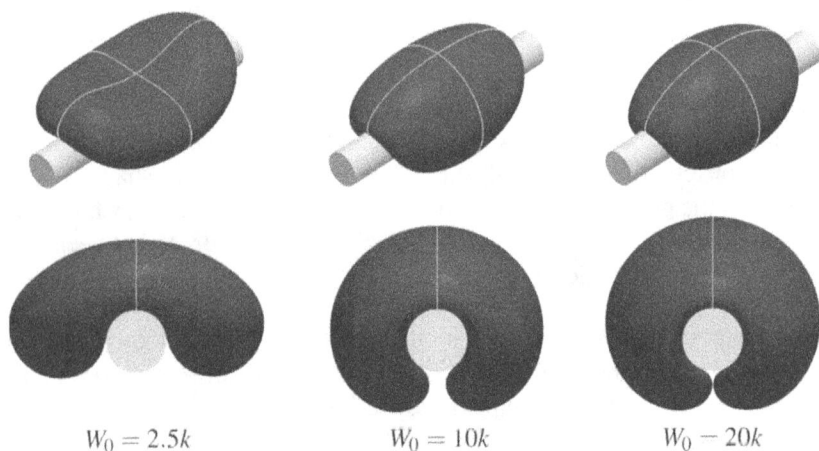

$$W_0 = 2.5k \qquad W_0 = 10k \qquad W_0 - 20k$$

Fig. 3.8. Computed RBCs shapes. The results of predictive mathematical modelling are based on Helfrich theory of membrane elasticity accounting for bending energy and cell-surface interaction W_0 (reprinted from [167], with permission).

3.3.3. QCM for the Studies of Clotting: Heparin Coating, Platelets Activation and Immunity Factors

In blood purification methods, clotting is a highly unwanted process. When blood cells are in the contact with biologically non-compatible surfaces, the adhesion of platelets, activation of the complement and the coagulation cascade can lead to the thrombosis as a complication [177, 178, 188]. Role of contact system activation in hemodialyzer-induced thrombosis has been studied in [178, 188].

Various surface coating techniques were employed to improve blood compatibility of polymer materials used in hemodialysis. Multilayer structures based on heparin immobilized on biological materials have been studied in [169]. In there, the QCM has been used for the real time studies of clotting processes [169].

Antithrombotic technologies for medical devices (see, for the latest reviews, in [170-178, 188]) include various methods of anticoagulation treatment of blood-contacting surfaces. Among them, the commonly used compound is a heparin covalently immobilized to the device surface [171]. The success in heparin coating allocation has been reported, for instance, in the artificial vessels research and for a new stent construction [172].

However, in hemodialysis, the correct usage of heparin is highly debated now. For example, some researchers argued that the heparin-coated filters for HD were not more effective than uncoated PS filters, [173]. Simultaneously, it was reported that oral anticoagulation treatment (5 % to 10 % of hemodialysis patients are treated with oral anticoagulants, according [174]) brought new challenges to be solved. For example, it was documented [174] that HD patients had an increased risk of bleeding caused by anticoagulation with heparin during hemodialysis. Also, it was stated that it is unknown whether additional anticoagulation with heparin or low-molecular-weight heparin is needed to prevent clotting during hemodialysis [174].

In this view, the additional detailed studies of the whole blood interaction with heparin compounds-either in solution or immobilized to the cells-sensors, with QCM-D and ECIS would be of great assistance. In particular, a strong motivation for the QCM/ECIS sensors improvement and further development is the search of new anticoagulant and antithrombotic compounds, evaluation of right dosage of the existing anticoagulants and a design of biologically tolerant coatings for blood contacting surfaces.

3.3.4. QCM for the Whole Blood Studies and Anticoagulants Research

The QCM-D technique has been successfully used in multiple studies of protein adsorption on surfaces (see, for the review, [91, 96]). A special attention has been focused on studies of fibrin adsorption since the activation of contact blood coagulation system may cause the formation of fibrin clots. Recently, the QCM-D has been employed for the study of complex protein-surface interactions such as complement activation system and contact activated blood coagulation [175] (see Fig. 3.9 and Fig. 3.10). In [169], the QCM-D has been used for real time measurements of blood coagulation density. The authors reported a comparative study of contact clot formation on different polymer surfaces.

Besides, the kinetics of fibrin adsorption on various polymer surfaces measured with the QCM [169,175] can provide the additional information on contact blood coagulation mechanism.

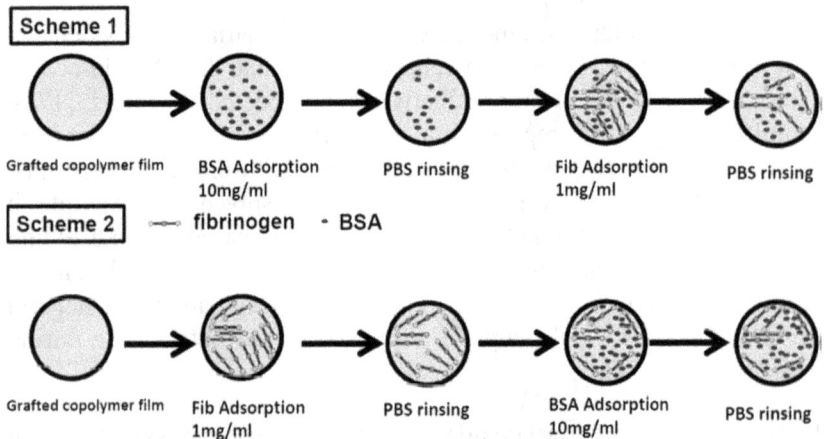

Scheme 1

Grafted copolymer film   BSA Adsorption   PBS rinsing   Fib Adsorption   PBS rinsing
                      10mg/ml   1mg/ml

Scheme 2   —— fibrinogen   - BSA

Grafted copolymer film   Fib Adsorption   PBS rinsing   BSA Adsorption   PBS rinsing
                      1mg/ml   10mg/ml

Fig. 3.9. 'Schematic illustration of fibrinogen adsorption on copolymer film (reproduced from [175], with permission).

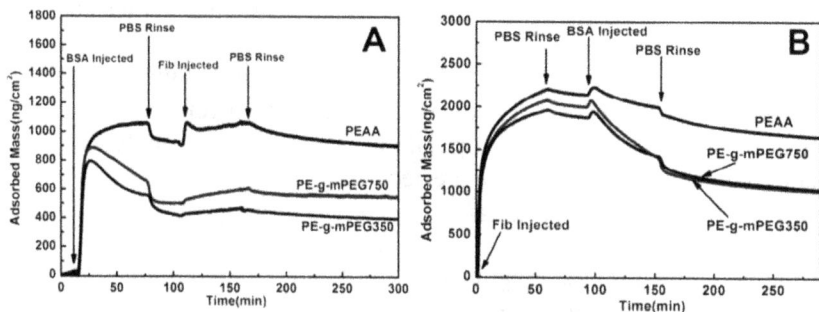

Fig. 3.10. The time course of the QCM resonance frequency changes during fibrin adsorption on different polymer surfaces in two dynamic schemes (A and B). Grafted copolymers: PEAA =Poly(ethylene-co-acrylic acid) (15% w/w acrylic acid) and poly(ethylene glycol) methyl ether (mPEG) (average Mw = 350 750). BSA=Bovine serum albumin, Fib=fibrinogen (from [175], with permission).

A new coating for QCM has been suggested in [176] for the monitoring of the activated partial thromboplastin time (aPTT) and fibrin adsorption. The effective adsorption of fibrin on highly hydrophobic Parylene (a transparent polymer widely used in medical devices) was found to enhance the signal-to-noise ratio.

In the papers [179-181], the changes in the QCM characteristics were related to the variations in fibrin viscoelasticity, i.e., in the loss and storage shear elastic moduli of protein layer adsorbed onto the QCM surface. The authors used a semi-phenomenological approach for the experimental data treatment. Thromboelastography (TEG) [182-184] is a classical method of aPPT measurement. A comparison of two methods, the standard (TEG) and QCM demonstrated a number of advantages for the latter technique [179].

Besides, several publications reported *in situ* QCM studies of blood coagulation parameters, such as prothrombin time (PT) or platelet aggregation [179, 180, 185-187]. In blood coagulation investigation, the comparative studies of QCM and coagulometer were performed in heparin- and heparin-less systems. In the paper [185], the authors make comment about QCM-based method as an alternative to a coagulometry. The researchers emphasized there [185], that the QCM method is different from standard coagulometry since it monitors the whole coagulation process in real time while the coagulometer measures the end point of PT only. In [185], a new affinity-based coating made of

91

polyethylene nanoparticles (PE-NPs) adsorbed onto a polymer-covered QCM surface has been reported. It was shown [185], that the sensitivity of PE-NPs-coated QCM was essentially higher. A comparison of the QCM data with mechanical coagulation (used as a reference method) demonstrated an excellent correlation. The authors concluded, that significant advantage of the suggested improvement was the reusability (up to 10 times) and robustness of the QCM coating [185].

The problems of blood coagulation disorders in chronic kidney disease have been recently reviewed in [188].

3.3.5. QCM for the Control of Hemodialysis Membranes Biofouling

Fouling of the filtration membranes [190-203] and the formation of biofilm are serious obstacles for long term utilization of equipment. In high dialysis performance, the dialysis fluid and blood flows should be uniform [191]. Biofouling restrict the flow across HD membranes and, consequently, reduce time of the dialysis filters usage. The adsorption of plasma proteins and adhesion of blood cells to the filters are the main concern in antifouling search. The correct evaluation of biofouling in hemodialyzer [192] and studying of mechanism of plugging and tortuous pores narrowing due to the plasma protein adsorption [193] is, thus, of great importance.

Since the fouling processes set a limit in multiple hemodialysis runs, the essential experimental efforts were directed to diminish clotting and the adsorption of proteins to HD filters. In particular, the surface treatment of HD filters includes the application of bovine serum albumin (BSE) [194], heparin coating [195, 196] and functionalization with lipids [197] and different polymers [198].

Studies of biofouling of porous filters with the QCM were recently reported elsewhere [199]. This work has been done with the aim to improve antifouling properties of the polysulfone (PSF), polyethersulfone (PES) and polyamide filter membranes, which are among the most widely used membrane filter materials in hemodialysis. Different methods were employed to avoid (or, at least, reduce) biofouling of membranes such as blending of polymers [199, 200], coating of filters with hydrophilic surfactants [201] and antithrombotic substances to prevent platelets adhesion [202]. Recently, the QCM usage for the optimization of antifouling coating has been reported in [203].

3.3.6. Theoretical Background for the Acoustic Biosensors (QCM and SAW-Based): Modelling the Response of Sensors in Measurements of Soft Biological Materials in Liquids

BAW and SAW sensors

Quartz crystal microbalance belongs to the class of bulk acoustic waves (BAW) acoustic devices. Another type of widely used acoustic sensors is based on surface acoustic waves (SAWs).

In general, the dynamics of polymeric and biomolecular materials can be characterized by measuring their viscoelastic response to oscillatory stress [204]. This is the underlying principle of acoustic sensors of both types in soft matter applications (for the review, see, e.g., [205, 206]). The analysis of the dispersion equations for BAW- and SAW-based devices provides mathematical expressions for the acoustic sensors characteristics, namely, the changes in the resonance frequency Δf and the dissipation factor ΔD for the QCM and the shift in SAW phase velocity, Δv, and the damping $\Delta \Gamma$, measured experimentally. Both BAW (QCM) [91, 92, 97, 100, 103, 105] and SAW-based [207-226] measurements reveal that biological structures cannot be considered as rigid films, but must rather be treated as soft (viscoelastic) materials. The variation in the environment (e.g., temperature, water content, ionic composition or addition of chemicals), interaction with the surface or biocomponents may significantly influence the softness of biological layers.

SAW-based sensors

In many biological applications, SAW-based devices are beneficial due to high operational frequency and, thus, their increased sensitivity [207-209, 211, 218, 226]. Surface acoustic waves with horizontal polarization (SH-SAW) -based devices are among popular methods [220-224] for the protein adsorption measurements and the detection of trace amount of additives in liquids [210, 211, 225, 227]. Since 90s, when the utility of SH SAW-based sensors in liquids was demonstrated for the first time [207-210, 213, 217, 218], these devices have become increasingly popular for the *in situ* measurements of biomolecular adsorption and evaluation of material properties of deposited biological layers (see, for the review, in [205, 211, 212, 215, 223, 224]).

Typically, in SH SAW devices, two measurable characteristics are the shift in phase velocity and the attenuation of acoustic waves caused by adsorbed layers [220-222]. These measurable can then be related to the mechanical properties of the layer, i.e. the surface mass and the viscoelastic parameters of the deposited material. It is well-known for the BAWs (QCM), that for the viscoelastic materials, the resonance frequency behavior deviates from Sauerbrey linear relationship due to the softness of material and viscous losses [91, 96-102, 107-114]. The theoretical work, thus, is vital for the analysis of soft deposited layers.

Below we provide a theoretical analysis for the surface acoustic waves propagating in the system of two soft biomolecular layers deposited onto the surface of SH SAW device (Fig. 3.11). This currently published work [115] is a generalization of viscoelastic membranes dynamics which was analyzed earlier in [113].

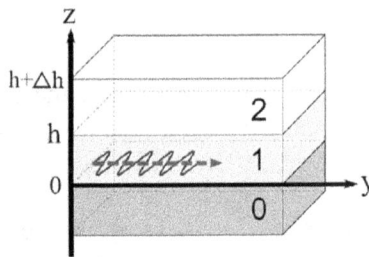

Fig. 3.11. Two-layers system (indices '1', '2') deposited onto the surface (index '0') of SH SSW-based sensor (from [115], with permission). Softness (viscoelasticity) of the layers is characterized by complex shear modulus of elasticity, $g^* = g' + ig''$, where g' and g'' are storage and loss moduli, respectively, i is the imaginary unit.

First, let consider a planar wave propagating in the y −direction (see the geometry, in Fig. 3.12) with the displacement u in the x-direction:

$$u = u_0 e^{\kappa z} e^{-iky + i\omega t} \tag{3.28}$$

with the general solution

$$u_i = u_0(A_i e^{\xi_i z} + B_i e^{\xi_i z}) e^{-iky + i\omega t} \quad (i = 1, 2) \tag{3.29}$$

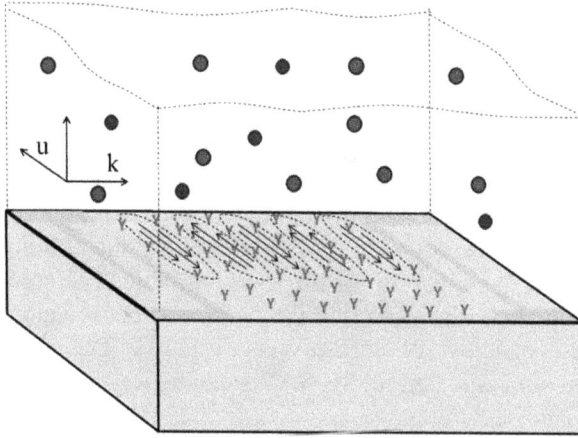

Fig. 3.12. Schematics of a SH SAW-based biosensor. Adsorption on the sensor's surface functionalized with SM (shown in blue) and receptors (shown in green) and loaded with a liquid on the top (from [114], with permission).

Here

$$\kappa = \sqrt{k^2 - \frac{\omega^2}{v_0^2}} \qquad (3.30)$$

is a wave vector (z-component) times the imaginary unit (the real part of κ is the inverse penetration depth into the bulk substrate), and the following notion is introduced:

$$\xi_i = \sqrt{k^2 - \omega^2 \frac{\rho_i}{g_i^*}} \ (i = 1, 2) \qquad (3.31)$$

Simultaneously, the substrate motion is described by the equation:

$$\rho_q \ddot{u}_y = g_q \left(\frac{\partial^2 u_y}{\partial z^2} + \frac{\partial^2 u_y}{\partial x^2} \right) \qquad (3.32)$$

Here, ρ_q and g_q are the density and shear modulus of elasticity of the substrate (index 'q' is for 'quartz').

By using the standard boundary conditions [228, 114, 115], together with no-slippage condition for the surface-adjacent layer, one can find the following equation [114, 115]:

$$\kappa = \frac{g_1^* \xi_1}{g_q} \frac{F_- - F_+ e^{2\xi_1 h}}{F_- + F_+ e^{2\xi_1 h}}, \tag{3.33}$$

where

$$F_\pm = g_1^* \xi_1 \pm g_2^* \xi_2 \tanh(\Delta h \xi_2) \tag{3.34}$$

Equation (3.33) together with the relations (3.29-3.32) and (3.33) gives an implicit dispersion equation for SH SAWs. By substituting k, ξ_i with κ, one can find the equation for the latter variable. The solution of the dispersion equation allows to find the change in phase velocity and the attenuation coefficient of surface waves [115]. The shift in phase velocity of a long-wavelength SH SAWs can be found as following:

$$\frac{\Delta v}{v_0} \approx -Re\left\{\frac{\kappa^2}{2(\frac{\omega}{v_0})^2}\right\} \tag{3.35}$$

and the attenuation coefficient (which is scaled by the wave vector k) is given by the expression:

$$\frac{\Gamma}{k} \approx -Im\left\{\frac{\kappa^2}{2(\frac{\omega}{v_0})^2}\right\} \tag{3.36}$$

where $v_0 = \sqrt{\frac{g_q}{\rho_q}}$ is a sound velocity in a quartz slab of density ρ_q and elasticity g_q.

The analytical formulae for the shift in phase velocity and the attenuation coefficient for the soft two-layer system are presented in [115].

By using equations (3.32-3.35), one can calculate phase velocity shift and the attenuation coefficient in different viscoelastic materials modelled in the Maxwell (viscoelastic fluid), Voigt (viscoelastic solid) or in a combined scheme of these two basic models.

In Fig. 3.14 one can find the results of computer calculations of the phase velocity changes and the attenuation coefficient for the SH SAWs based on the above mentioned model (see equations (3.33-3.36)). The plots were calculated for different parameters of the layer, namely, η_1, the shear viscosity, and μ_1, the modulus of shear elasticity (In Voigt scheme, $g^* = g' + ig''$; $g' = \mu$, $g'' = \eta \omega$).

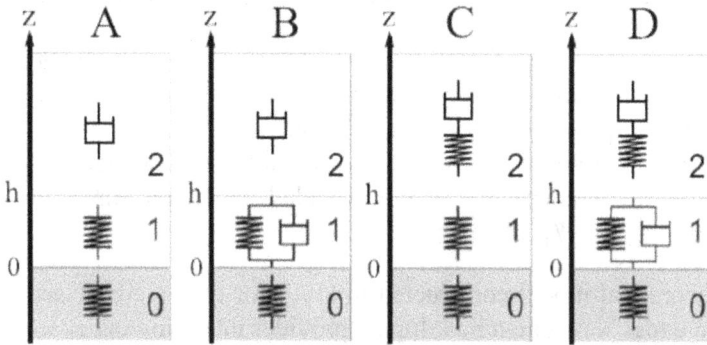

Fig. 3.13. Four different viscoelastic schemes for the two-layers-on-substrate system. The substrate (a quartz crystal, index '0') is modelled as a pure elastic element (a spring), while layers are represented as a combination of elastic (a spring) and viscous (a dashpot) elements. The Maxwell material is represented with spring and dashpot elements in series, while Voigt-type material is a combination of elastic and viscous elements in parallel (reproduced from [115], with permission).

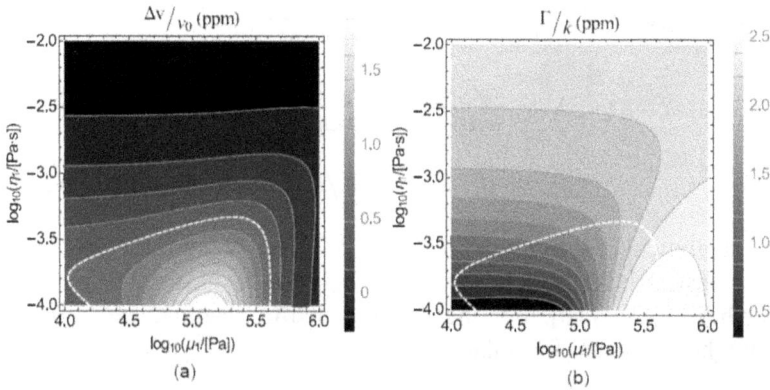

Fig. 3.14. The shift in the phase velocity (a) and the scaled attenuation coefficient (b) for a viscoelastic solid (Voigt scheme) film under bulk Newtonian liquid calculated at 100 MHz (from [115], with permission).

The unknown viscoelastic parameters of the layer can be determined via simultaneous measurements of the phase velocity shift and the attenuation coefficient if the material parameters of the layer, i.e., its density and thickness is known or can be found by other method(s).

97

Measurements of viscoelastic layer parameters with QCM-D

In a similar way, it was found in case of BAW sensors and, in particular, for the QCM-D, that the simultaneous measurements of the shift in the resonance frequency of the sensor, Δf, and the dissipation, D, may bring the information about the viscoelastic moduli of the material, g', g''. The interested reader can find the extended review on this subject, for example, in [91, 97].

The theoretical model constructed in [109] for the layered viscoelastic system under Newtonian bulk liquid, provides the following expressions for the resonance frequency shift and the dissipation of the QCM:

$$\Delta f = \frac{Im(B)}{2\pi m_q}, \Delta D = -\frac{Re(B)}{\pi f m_q}, \tag{3.37}$$

where

$$B = \kappa_1 \xi_1 \frac{1 - Ae^{2\xi_1 h_1}}{1 + Ae^{2\xi_1 h_1}} \tag{3.38}$$

$$A = \frac{\kappa_1 \xi_1 (1 + \Lambda e^{2\xi_1 \Delta h_1}) - \kappa_2 \xi_2 (1 - \Lambda e^{2\xi_1 \Delta h_1})}{\kappa_1 \xi_1 (1 + \Lambda e^{2\xi_1 \Delta h_1}) + \kappa_2 \xi_2 (1 - \Lambda e^{2\xi_1 \Delta h_1})} \tag{3.39}$$

$$\Lambda = \frac{\kappa_2 \xi_2 + \kappa_3 \tanh(\xi_3 \Delta h_2)}{\kappa_2 \xi_2 - \kappa_3 \tanh(\xi_3 \Delta h_2)} \tag{3.40}$$

$$\xi = ik; \kappa = -iG/\omega \tag{3.41}$$

The analytical expressions for the Δf and D in case of a soft (viscoelastic) layer under Newtonian bulk liquid have been published elsewhere [109, 110]. For the convenience of reading, we reproduce below the analytical formulae:

$$\Delta f \approx -\frac{\eta_L}{2\pi m_q} - \frac{h\rho\omega}{2\pi m_q}\left\{1 - \frac{2}{\rho}\left(\frac{\eta_L}{\delta_L}\right)^2 J''\right\} \tag{3.42}$$

$$\Delta D \approx \frac{1}{\pi f m_q}\left\{\frac{\eta_L}{\delta_L} + 2h\omega\left(\frac{\eta_L}{\delta_L}\right)^2 J'\right\} \tag{3.43}$$

In the expressions (3.42, 3.43), δ is the viscous penetration depth, i.e. the distance over which the transverse wave amplitude falls off by a factor of e, index 'L'' denotes bulk liquid, J' and J'' is the compliance of the soft (viscoelastic) film (the ratio of the storage and loss components):

$$\delta = \sqrt{2\eta_L/\rho_L\omega} \qquad (3.44)$$

$$J' = \frac{g'}{g'^2 + g''^2} \qquad (3.45)$$

$$J'' = \frac{g''}{g'^2 + g''^2} \qquad (3.46)$$

Note that the expression (3.42) for the resonance frequency shift consists of one bulk term and two terms proportional to the film surface mass. The latter are opposite in sign: this finding of the mass correction is due to the presence of viscous liquid on the top of the film (the so-called "missing mass effect", [110]). Recently, the similar effect has been revealed for the SH SAW case in [114, 115]. The "missing mass" effect is a consequence of the interplay between the viscoelasticity of the overlayer and that of a top fluid, in both, BAW (QCM) and SAW (SH SAW) situations. In both cases, the dispersion equation allows to derive the analytical expressions for the experimentally measured characteristics, the shift in the resonance frequency (or phase velocity) and the dissipation (attenuation). These expressions are functions of material parameters of the layer(s). Numerical simulations performed in [109, 110, 115] for selected examples of viscoelastic materials confirmed the analytical results and illustrate the 'missing mass' effect, a correction to the QCM resonance frequency shift (or phase velocity in case of SH SAW) and the dissipation (attenuation) due to viscoelastic coupling between the overlayers [114, 115].

The analytical expressions for the experimentally verified BAW and SAW acoustic sensors reviewed in the present section, are useful theoretical tool for the quantitative interpretation of QCM and SH-SAW results in biological and biomedical applications.

The change in the resonance frequency of QCM coated with a soft layer (taken as a Voigt viscoelastic solid material) of thickness h and of density ρ under a Newtonian liquid (considered as a bulk medium) [109, 110]:

$$\Delta f \approx -\frac{1}{2\pi m_q}\left\{\frac{\eta_L}{\delta_L} + h\rho\omega - 2h\left(\frac{\eta_L}{\delta_L}\right)^2\frac{\eta\omega^2}{\mu^2 + \eta^2\omega^2}\right\} \qquad (3.47)$$

The dissipation factor of the QCM:

$$\Delta D \approx \frac{1}{\pi f m_q} \left\{ \frac{\eta_L}{\delta_L} + 2h \left(\frac{\eta_L}{\delta_L} \right)^2 \frac{\mu \omega}{\mu^2 + \eta^2 \omega^2} \right\} \qquad (3.48)$$

The expressions (3.47, 3.48) are obtained in the linear approximation when the film thickness h is taken as a small parameter compared to the viscous penetration depth δ_L in liquid.

The different approach for the theoretical analysis has been suggested in [97-99]. The equivalency of the results of both theoretical models has been demonstrated in [97] (see, Appendix in there). Also, a comparison of formulae (3.47, 3.48) obtained within the continuum approach [109, 110] and those calculated within the transmission-line equivalent-circuit model in [102], shows that in the case of a QCM coated with a thin viscoelastic layer and operated in a Newtonian liquid the results are identical. In recently published paper [105], Kanazawa, Frank and Hardesty compared the QCM experimentally measured characteristics with the results of theory. The authors [105] demonstrated that their results and of two above mentioned models are indistinguishable.

Recently published work [273] represents general theoretical results, including analytical formulae for the resonance frequency changes and the dissipation of QCM-D, for two viscoelastic layers deposited onto resonator surface, as well as calculations for higher harmonics. In particular, the corrections due to the softness of the top layer, considered as a bulk, are written explicitly in terms of the layer's material parameters [273].

3.3.7. QCM-EIS/ECIS Combined Sensing

One of the most promising direction in biosensors improvement is the integration of quartz crystal resonator (QCM) with electrical impedance spectroscopy (EIS) [229]. Quartz crystal microbalance and impedance spectroscopy are named as main developments of last two decades in non-invasive characterization of biomolecular materials deposited onto the sensor's surface [230]. In the past configurations, the QCM top electrode was used as the working electrode in EIS measurements while the second counter electrode was dipped into solution and the electrical impedance was measured between them [229]. The crucial improvement in engineering of transducers was made by Janshoff and Wegener and reported in [231]. In this work, the double-mode of impedance analysis was demonstrated for the first time in biosensors studies of adsorbed

cells. The additional low impedance platinum electrode was introduced
into the system to measure simultaneously EIS- and QCM-mode. This
combined double-mode approach was applied for measurements of cells
cultivated directly on the QCM surface [231]. In this work, a
double-mode impedance analysis has been applied to epithelial cell
monolayers cultured on acoustic resonator surface [231].

The study of living cells electrochemical impedance started from pioneer
work of Giaever and Keese [232], where the EIS measurements were
used for recording and analysing the impedance of cells deposited on
gold electrodes. This method is referred now as electrical cell-substrate
sensing (ECIS) [229]. In ECIS, the properties of cell layer on electrode
and the electrolyte are represented as an equivalent electrical circuit with
resistive and capacitive elements connected in series or in parallel [233,
234]. Experimentally measured electrical impedance magnitude and
phase angle provide valuable data for characterization of cell
membranes. The advanced impedance spectroscopy analysis of living
cells sensitivity and frequency response measured with ECIS method has
been reported elsewhere (see, for the review, [234] and references
in there).

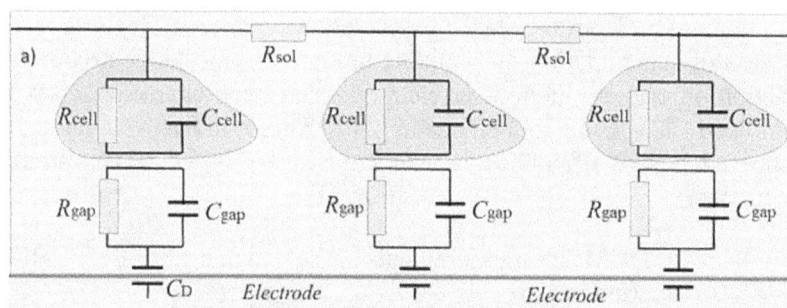

Fig. 3.15. The equivalent electrical circuit of the ECIS-sensor coated with cells
layer.

Current progress in this field includes a design of sensors for
simultaneous viscoelastic and dielectric measurements of cells in
combined QCM and EIS/ECIS studies (see, for the review, in [229]).
These devices are aimed to collect the complementary information about
complex biological systems such as measurement of cell membranes
mechanical and electrical properties, cell adhesion and cell-surface

interaction [229, 230], the cells functioning including the activity of ion channels [230].

There is a long list of achievements in this field and a complete review of QCM/EIS measurements in biosensors is beyond the limits of current work. For the interested readers, one should recommend an extended topical description given in [229] and [230]. Below we give just a brief account of some results in this area.

QCM combined with EIS method has been used in [235] for simultaneous impedance measurements of two quartz crystals in solution. These combined methods were used to monitor *in situ* the antibody adsorption and specific antibody-antigen reaction. For example, the QCM (Δf, the resonance frequency shift) and EIS (ΔC, the interfacial capacitance) characteristic measured simultaneously have been utilized to estimate the immunoreaction parameters and to study proteins (antihuman immunoglobulin G and human immunoglobulin G) adsorption on bare electroplated QCM gold electrodes in [236]. Bovine serum albumin (BSA) adsorption on modified QCM electrode was reported in [237, 238]. The interaction of BSA protein with metals (copper) was studied with electrochemical impedance spectroscopy and QCM in [239].

For studies of electron-transfer processes in enzymatic electrodes, the process of accumulation of insoluble product of biocatalized oxidative reaction on enzyme-monolayer electrodes has been characterized with combined QCM and EIS methods in [240]. These studies were aimed to improved design of bioelectrochemical sensors based on enzymatic electrodes.

Simultaneous QCM/EIS measurements of the polymer films degradation have been reported in [241].

The supported lipid bilayer formation and interaction with pore-forming peptides have been investigated by using the combined QCM-D and EIS methods in [242].

Some other examples of the successful application of combined QCM-EIS/ECIS method are given in the next section on biosensors for water quality control.

3.3.8. QCM, ECIS or QCM Combined with EIS/ECIS in Water Control for Dialysis

3.3.8.1. Importance of Water Quality Control for HD

The high quality of water is crucial for hemodialysis. In most cases, a drinking water is used in the water treatment system for the preparation of dialysate. However, even at low concentration, a presence of pollutants in a tap water can be harmful to HD patients and cause a severe health problem if these substances penetrate through water filters appearing in dialysate. Safety in dialysis demands that water used for the dialysate preparation must be checked for the presence of toxic additives and thoroughly purified to satisfy the criteria of worldwide standards for HD water treatment [243-245].

Existing criteria for the dialysate composition [245] may vary from country to country. World standards for dialysis water quality set a different limit for substances normally included in dialysate for calcium, magnesium, potassium and sodium [243-245].

However, there is a general international agreement that heavy metal contaminants are toxic to hemodialysis patients if appeared in dialysate [244, 245]. Among inorganic substances, mercury Hg^{2+} ions are considered as the most harmful [244]. Other toxic for HD patients metal ions are aluminum, zinc and copper [245]. Some of the organic pollutants coming from water supply to a tap water may contain also bacterial toxins. These toxins are much smaller than bacteria and can contribute to water pollution [244]. Thus, avoiding toxicity in a dialysate requires an effective control over the composition of drinking water, water-borne contaminants including monitoring of water supply's conditions. The recognition of the problem of ultrapure water for dialysate requires a further progress in sensor technologies for the detection and real time monitoring of extremely low concentration of water pollutants [243-247].

In this section we give a short review on application of QCM, EIS/ECIS and combination QCM-EIS methods for the detection of heavy metal ions and bacterial toxins in water.

3.3.8.2. Designing Biosensors for the Detection of Heavy Metals Ions in Solution

EIS/ECIS -based sensors

Two metals-copper and mercury ions-are the special targets for the clean water sensors.

Mercury ion (Hg2 +) is a highly toxic water soluble pollutant [153, 248-255]. Since mercury ions can come to the human not only with water but also via food chains, the development of Hg + 2 sensors is of great importance in a food quality control, as well as in environmental and biomedical applications (in particular, in toxicology and water treatment for HD).

A number of techniques have been suggested for the detection of mercury ions. Among them, one should mention a usage of semiconductor quantum dots [253], carbon-based fluorescence [254, 255], fluorescence resonance energy transfer (FRET) detection by polymer-based nanoparticles [256], colorimetric gold nanoparticles [257] and SERS (of metal nanoparticles) [258] methods.

However, the bottle neck in the above mentioned methods is their low sensitivity. Going in biosensors' direction, recently a fabrication of metal ions sensing DNA-in combination with enzymes-platform has been reported elsewhere [153]. In biosensors' configuration, the coordination chemistry of an enzyme has been used in ion-dependent DNAzyme response [153]. The authors [153] reported the results of electrochemical impedance (EIS) measurements by using the hybrid DNAzyme biosensor. They demonstrated a high sensitivity of its EIS characteristics to Hg2 + ions measured in the increased range of mercury ions concentration in solution.

Another type of impedimetric DNA-based sensor for Hg2 + detection has been designed in [259]. It was shown that the conformational changes of DNA induced by heavy metal ions (Hg2 + and Pb2 +) led to a decreased R_{CT} which was tracked with electrochemical impedance spectroscopy (EIS) method.

DNA-based biosensors

A systematic investigation of charge transfer (CT) in DNA in electrochemical sensors started in 1990s. In the paper by Kelly and Barton [260], the ultrafast dynamics of CT through DNA over a large distance was revealed. Also, it was found that the redox reactions in solution were sensitive to sequence and dependent on base-base interactions.

The crucial step in developments of DNA-based biosensors was the discovery of specific interaction of heavy metals ions (such as Hg2 +, Ag + [261] and Pb2 + [250, 259, 262]) and base pair mismatches: C-C (for the silver ions) [261] and T-T mismatches (in case of Hg2 +) in DNA [249].

The silver ions (Ag +) in water are sometimes not treated as harmful components. In the opposite, this metal is often considered as a healthy additive. However, it is documented that at concentrations above the certain level, the presence of Ag + ions in drinking water or water used for the dialysate could be dangerous for the patients [263]. Owing to the strong oxidation of Ag + properties [263], when coming into the blood, these heavy metal ions may induce severe complications such as internal organs edema [263]. Thus, the development of DNA-based sensors for silver ions detection in dialysate water has the same high importance as for the detection of mercury or copper ions [263]. Since in real water samples (of drinking water or lake water) there are multiple ions, the simultaneous and selective detection of heavy metals traces performed in one sensor device is of a great practical need [261].

DNA-based biosensors have attracted attention due to their high sensitivity and the ability to detect biomolecular interactions with a single target or even multiple target analytes [258, 259, 264] such as proteins and toxic metal traces. An example of a parallel detection of proteins (thrombin) and metal ions in a multifunctional DNA-based biosensor has been demonstrated in [264]. In this work, the electrodes modified with DNA recognizing mercury ions have been used to measure the difference in R_{CT} against changes in Hg2 + concentration in water.

Recently, a number of important publications on DNA-based sensors for the analysis of water containing a mixture of heavy metals ions has been published elsewhere (see, for the review [249] and references in there). For example, in [249], the impedimetric DNA-based sensors for a simultaneous detection of Pb2 +, Ag + and Hg2 + ions are reported. The schematic depiction of the sensor based on DNA platform is shown on Fig. 3.16.

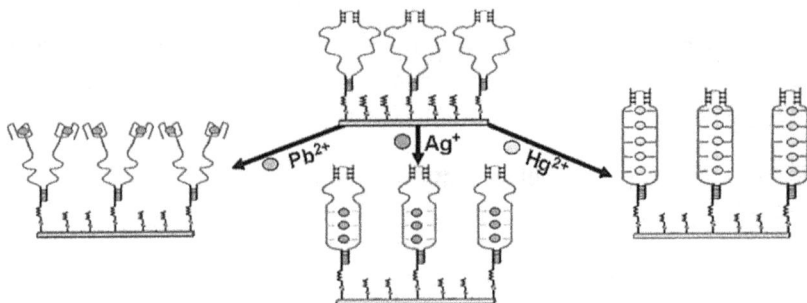

Fig. 3.16. Sketch of the DNA-based platform for a simultaneous detection of Pb2 +, Ag2 + and Hg2 + ions in solution (reprinted from [249], with permission).

For the impedimetric studies, the structural components of the biosensor are represented in the form of an equivalent electrical circuit consisting of resistive and capacitive elements (see Fig. 3.17). The most important electrical characteristic, R_{CT}, is the resistance due to the charge transfer (CT) between the redox probe in solution and the electrode surface [249].

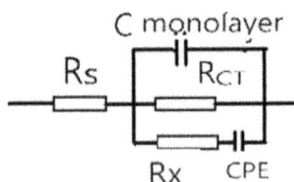

Fig. 3.17. The sketch of the equivalent electrical circuit for the DNA-based biosensor shown on Fig. 3.16.

In Fig. 3.17, $C_{monolayer}$ denotes the capacitance of DNA films on gold electrodes, R_S corresponds to the resistance of solution, the combination of R_X and CPE (constant phase element) describes electrical properties of the surface-modified electrodes (known). The electrically measured parameter is the difference between R_{CT} value and the resistance of solution, $\Delta R_{CT} = R_{CT} - R_S$. The experimentally observed decrease in R_{CT} value has been explained by authors as a formation of T- Hg2 + -T base pairs and enhanced electron transfer [249].

Interesting variant of DNA-based sensor combined with the redox enzyme has been developed in [265]. In this work, a sensor bioelectronic platform of DNA template nanowired to the gold electrode has been engineered. A nucleic acid- modified glucose oxidase (GOx) enzyme was hybridized via a cooperative bridging with T-Hg2 + -T pairs of the duplex. The charge transfer in the system was mediated by a redox unit (ferrocene molecule tethered to the gold electrode surface).

Monitoring heavy metal ions with QCM sensors

Molecular imprinting (MIP) technology is currently employed to design polymer templates with special recognition sites for certain compounds. In particular, a modification of Cu(II) ion-imprinted polymer (Cu(II)-IIP) film has been used in combination with QCM for the detection of Cu(II) ions in solution in [266]. Polymer -grafted QCM sensor has been developed and applied to heavy metal ions monitoring in [267]. The polymer-grafter films onto the QCM surface was found can selectively adsorb heavy metal ions (copper, cadmium, lead and chromium) from solution by forming complexes with polymer functional groups.

Similar idea of sensor surface coating with a specific chemical response to heavy metals has been realized in [268]. In this work, the QCM surface was functionalized with a phosphate-modified dendrimer film. The ability of polymer to coordinate metal ions in their branches makes a principle of the selectivity of the sensor.

3.3.8.3. Biosensors for Label-Free Sensing of Water Contaminants (Based on QCM and EIS/ECIS)

Cyanobacterial toxins analysis in water

Toxins produced by cyanobacteria can penetrate across filters and pollute both drinking water and dialysate [243]. Such cyanobacterial toxins as microcystins (MC) are found to be harmful to humans and animals [269] being promoters for acute hepatotoxicity. The violation of world water standards for HD [244, 245] and careless attitude to dialysis facilities may have dangerous consequences for the patients and contribute to long-term morbidity and health damage [243].

Contamination of drinking water (as a main source of water for dialysate preparation) with cyanotoxins is currently measured with a number of different methods [269-271]. The traditional (biological) methods

[268-271], such as ELISA, mass spectrometry and HPLC are robust but require the expensive equipment and not able to measure the pollution in real time. Another threat is the formation of biofilm in the HD equipment. The control over both processes demands *in situ* and real time monitoring in the dynamic environment involved in water treatment, dialysate preparation and biofilm formation in water pathways in dialysis equipment.

The dialysis water treatment can be damaged with unexpected rise in bacterial contamination due to, for example, seasonal variations in the surface water resources (rivers, lakes) during the period of algal bloom [243-245]. Another problem is a diversity of biomolecular structure and modes of action of bacterial toxins. The striking example provides a chemical variation in the MCs species [270]. More than 80 variants of the MC toxins different in their amino acids position are documented. One of the most studied toxin is known as microcystin-LR (MC-LR) [270-272].

To summarize all above mentioned, there is a great need in sensor devices to detect and monitor in real time the microbial toxins that may potentially contaminate a fresh water used for dialysis. These devices should be highly selective on molecular level to the specific microcystin toxin. Besides, such biosensors should satisfy the criteria of portability and (in the ideal situation) low cost.

QCM sensor for bacterial toxins monitoring

Quartz crystal microbalance (QCM) and a quartz crystal microbalance with dissipation monitoring (QCM-D) is a popular method for in situ detection of contaminants in liquid environments. A combination of QCM with microfluidics, for example, in a Lab-on-a-chip combined scheme [271], is an excellent tool for label-free sensing of water toxicants such as MCs.

Another promising system for the enhancing MCs detection is a new immunosensor reported in [272]. In this work, the sensitivity and specificity of immunoassays (similar to ELISA) were combined with advantages of QCM such as portability, small size of the device and cost effectiveness due to minimal reagents needed. To make the sensor selective to the specific target, the QCM was coated with the layer of monoclonal antibodies against MC-LR [272] by using the surface adsorption of lipid vesicles.

Sensing water toxins with EIS/ECIS and in combination with QCM method

Sensors based on ECIS method suggested in [232] have been successively applied for the toxicology studies due to the easy operation [271] and a high sensitivity of various kinds of cultured living cells to the toxins in water. In this approach, the living cells *per se* are used as a sensing element. When the cells were exposed to biological contaminants, their physiological behaviour changed. This biological response was then converted into electrical signals and monitored as an electrical impedance. Living cells as toxicity sensors may thus indicate a presence of a target analyte or alarming a harmful effect of the unknown compound present in a tested solution [271].

A miniaturization is an important aspect in immunosensors design. Recently, a number of technical solutions and microfluidic devices combining a fluidic cell with ECIS impedimetric analysis has been published elsewhere (see, for the review, in [271]). The innovative design of multifunctional sensor has been reported in [271]. In this work, the electrical cell-substrate impedance sensing (ECIS) method was integrated with the QCM measurements. The authors emphasized a high accuracy of the toxin's detection and increased sensitivity of this combined method in water tests [271].

The change in the electrical impedance and in the resonance frequency (and the dissipation in the QCM-D devices) have been monitored simultaneously in response to the cell damage or death due to the toxicants in water [271] (Fig. 3.17). In this study, a reduction in the ECIS impedance has been interpreted as related to the cells apoptosis and death caused by the presence of toxicants in water. Simultaneously, the resonance frequency of the QCM increased that was explained as a detachment of the apoptotic cells from the QCM surface.

3.4. Outlook

Theoretical physics is aimed to uncover existing relations between various natural phenomena, at first sight very different, and formalize them in mathematical equations. Maxwell-Wagner-Hanai theory of the interface polarization in mixtures (the "mixture model") provides an excellent example of one of the most successful physico-mathematical models with multiple practical applications. In particular, this theory lied a fundamental basis for the electrical impedance and impedance spectroscopy analysis (BIA/BIS) widely used in biological and medical studies related to blood purification. The remarkable feature of this theory is its universality; it works on different levels and scales,-from bioimpedance measurements of the whole organism (i.e., in BIA for the evaluation of body composition, TBW, ICW and ECW) to tissues and individual living cells studies (i.e., in EIS/ECIS).

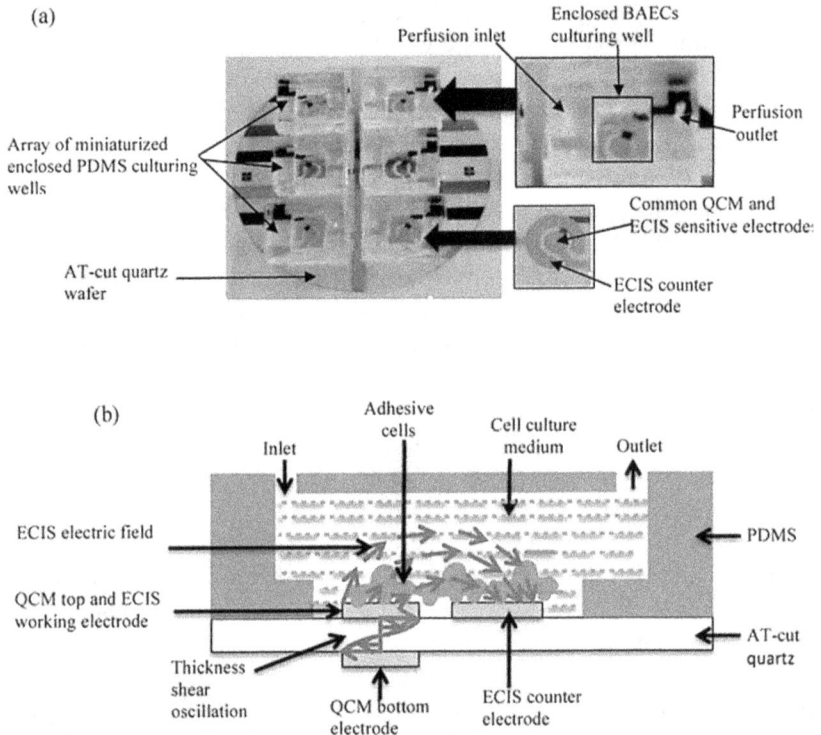

Fig. 3.17. Combined QCM and ECIS microfluidic device. The sensor is based on measurement of cultivated bovine aortic endothelial cells (BAECs) response, living sensors of toxins in fluid medium (reproduced from [271], with permission).

The continuous challenge for both theory and engineering work is the optimization and control of the hemodialysis process aimed at the individual patient healthcare. At the moment, the complete solution remains to be found and realized in clinical practice with the help of HD controlling bioelectronics systems. Despite of the successful application of impedance spectroscopy for these purposes, there are several bottlenecks, such as the difficulties in distinguishing between body fluids during the process of hemodialysis and, consequently, to find the right dose of the fluid amount and composition to be introduced (which is different to each patient) to the organism (the "dry weight" problem). Going in this direction, the search of compromise between the sensitivity of devices and low cost robust solutions is crucial for the everyday usage of electronics in HD clinical practice.

The effectiveness, safety and long term exploitation of the HD equipment is another important issue. In there, the coagulation of blood and biofouling of HD membrane filters are among the top listed problems. The search of most effective anticoagulant and antifouling coatings of the membrane filters demands new tools which unite electronic and sensor components in multifunctional biosensor devices. The combination of impedimetric (BIA/EIS) and acoustic (quartz crystal microbalance, QCM) devices in one hybrid sensor system is a noteworthy example of a new promising technological solution. In the present review we demonstrated the successful applications of EIS and acoustic sensors (QCM and SAW-based) and combined QCM/EIS methods in the search of new coatings for the HD filters as well as in associated analyses of plasma proteins and blood cells contacting these filters. One should note, however, that usage of microscopy and other (reference) methods could be very helpful, in particular, for the calibration of devices.

Theoretical background is significant for the correct interpretation of experimentally measured characteristics. For this reason, a special section of the current review is dedicated to the theory behind acoustic sensors of QCM and SAW-based type, when operated in liquids and measured soft biological materials.

Finally, the safety in dialysis is strongly dependent on quality of water used for dialysate. We brought this important aspect for the readers' attention. Since even a trace amount of heavy metal ions or other toxins in water can lead to severe health damage and morbidity of HD patients, the sensors' analysis of dialysate water sources is a key factor. Due to possible variations in water composition, it is crucial to monitor water quality in real time. The acoustic sensors (both of QCM and SAW-based types) and a combination of them with impedimetric devices in an electronics platform or L-O-C microfluidic system is a precise analytical tool for real time measurements of additives in water. In the review we present the R&D efforts in this life important topic.

Acknowledgements

One of the authors (MV) is grateful to Dr. Bogdan Tkachuk who pointed out the importance of high standards for water quality used in dialysate preparation and for the fruitful discussions of impedance measurements in hemodialysis.

References

[1]. J. Ahlmen, Quality of Life of The Dialysis Patient, Replacement of Renal Function by Dialysis, 5th Ed., Vol. 3, *Springer*, 2004, pp. 1315-1332.

[2]. S. Schmaldienst, W. H. Hörl, The Biology of Hemodialysis, Replacement of Renal Function by Dialysis, Vol. 1, *Springer*, 2004, pp. 157-179.

[3]. Progress in Hemodialysis-From Emergent Biotechnology to Clinical Practice (A. Carpi, C. Donadio, G. Tramonti, Eds.), *InTech*, 2011.

[4]. Biofeedback Systems and Soft Computing Techniques of Dialysis, Modelling and Control of Dialysis Systems (A. T. Azar, Ed.), Vol. 2, *Springer*, 2013.

[5]. M. C. Flessner, Kinetic modelling in peritoneal dialysis, in Biofeedback Systems and Soft Computing Techniques of Dialysis, Modelling and Control of Dialysis Systems (A. T. Azar, Ed.), Vol. 2, *Springer*, 2013, pp. 1427-1476.

[6]. U. G. Kyle, I. Bosaeus, A. D. De Lorenzo, P. Deurenberg, M. Elia, J. M. Gomeze, B. L. Heitmann, L. Kent-Smitht, J. C. Melchior, M. Pirlich, H. Scharfetter, A. M. Schols, C. Pichard, Bioelectrical impedance analysis-part I: Review of principles and methods, *European Journal of Clinical Nutrition*, Vol. 23, 2004, pp. 1226-1243.

[7]. U. G. Kyle, I. Bosaeus, A. D. De Lorenzo, P. Deurenberg, M. Elia, J. M. Gomeze, B. L. Heitmann, L. Kent-Smitht, J. C. Melchior, M. Pirlich, H. Scharfetter, A. M. Schols, C. Pichard, Bioelectrical impedance analysis-part II: Utilization in clinical practice, *European Journal of Clinical Nutrition*, Vol. 23, 2004, pp. 1430-1453.

[8]. S. J. Davies, A. Davenport, The role of bioimpedance and biomarkers in helping to aid clinical decision-making of volume assessments in dialysis patients, *Kidney International*, Vol. 86, 2014, pp. 489-496.

[9]. S. Grimnes, Ø. G. Martinssen. Bioimpedance and Electricity Basics, *Academic Press*, 2000.

[10]. M. Hecking, A. Karaboyas, M. Antlanger, R. Saran, V. Wizemann, C. Chazot, H. Rayner, W. Hörl, R. L. Pisoni, B. M. Robinson, G. Sunder-Plassmann, U. Moissl, P. Kotanko, N. W. Levin,

M. D. Säemann, K. Kalantar-Zadeh, F. K. Port, P. Wabel, Significance of interdialytic weight gain versus chronic volume overload: consensus opinion, *American Journal of Nephrology*, Vol. 38, 2013, pp. 78-90.

[11]. M. Y. Jaffrin, H. Morel, Body fluid volumes measurements by impedance: A review of bioimpedance spectroscopy (BIS) and bioimpedance analysis (BIA) methods, *Medical Engineering & Physics*, Vol. 30, 2008, pp. 1257-1269.

[12]. M. Y. Jaffrin, M. Fenech, M. V. Moreno, R. Kieffer, Total body water measurement by a modification of the bioimpédance spectroscopy method, *Medical & Biological Engineering & Computing*, Vol. 44, 2006, pp. 873-882.

[13]. J. R. Matthie, Second generation mixture theory equation for estimating intracellular water using bioimpedance spectroscopy, *Journal of Applied Physiology*, Vol. 99, 2005, pp. 780-781.

[14]. J. R. Matthie, Bioimpedance measurements of human body composition: critical analysis and outlook, *Expert Review of Medical Devices*, Vol. 5, 2008, pp. 239-261.

[15]. A. De Lorenzo, A. Andreoli, J. Matthie, P. Withers, Predicting body cell mass with bioimpedance by using theoretical methods: A technological review, *Journal of Applied Physiology*, Vol. 82, 1997, pp. 1542-1558.

[16]. H. C. Lukaski. Bioelectrical impedance analysis, in *Proceedings of the AIN Symposium (Nutrition '87)*, Bethesda, 1987, pp. 78-81.

[17]. H. C. Lukaski, Methods for the assessment of human body composition: traditional and new, *Journal of the American College of Nutrition*, Vol. 46, 1987, pp. 537-556.

[18]. H. C. Lukaski, W. W. Bolonchuk, Estimation of body fluid volumes using tetrapolar bioelectrical impedance measurements, *Aviation, Space, and Environmental Medicine*, Vol. 59, 1988, pp. 1163-1169.

[19]. R. F. Kushner, Bioelectrical impedance analysis: A review of principles and applications, *Journal of the American College of Nutrition*, Vol. 11, 1992, pp. 199-209.

[20]. R. F. Kushner, D. A. Schoeller, Estimation of total body water by electrical impedance analysis, *Journal of the American College of Nutrition*, Vol. 11, 1986, pp. 417-424.

[21]. S. S. Sun, W. C. Chumlea, S. B. Heymsfield, H. C. Lukaski, D. Schoeller, K. Friedl, R. J. Kuczmarski, K. M. Flegal, C. L. Johnson, V. S. Hubbard, Development of bioelectric impedance analysis prediction equations for body composition with the use of a multicomponent model for use in epidemiologic surveys, *American Journal of Clinical Nutrition*, Vol. 77, 2003, pp. 331-340.

[22]. S. B. Heymsfield, Z. M. Wang, R. T. Withers, Multicomponent molecular level models of body composition analysis, in Human Body Composition: Methods and Findings (A. F. Roche, S. B. Heymsfield, T. G. Lohman, Eds.), *Human Kinetics*, Champaign, IL, 1996, pp. 129-148.

[23]. J. C. Maxwell, A Treatise on Electricity and Magnetism, 2nd Ed., *Clarendon Press,* Oxford, 1881.

[24]. K. W. Wagner, Erklärung der dielektrischen Nachwirkungsvorgänge auf Grund Maxwellischer Vorstellungen, *Arch. Electrotechn.,* Vol. 2, 1914, pp. 371-387.

[25]. D. A. G. Bruggeman, Berechnung verschiedener physikalischer Konstanten von heterogenen Substanzen. I. Dielektrizitätskonstanten und Leitfähigkeiten der Mischkörper aus isotropen Substanzen, *Annalen der Physik,* Vol. 416, 1935, pp. 636-664.

[26]. R. E. De la Rue, C. W. Tobias, On the conductivity of dispersions, *Journal of Electrochemical Society,* 1959, pp. 827-833.

[27]. T. Hanai, Theory of the dielectric dispersion due to the interfacial polarization and its application to emulsions, *Kolloid-Zeitschrift und Zeitschriftb fur Polymere (Colloid and Polymer Science),* Vol. 171, 1960, pp. 23-31.

[28]. T. Hanai, N. Koizumi, Dielectric relaxation of W/O emulsions in particular reference to theories of interfacial polarization, *Bulletin of the Institute for Chemical Research,* Kyoto University, Vol. 53, 1975, pp. 153-160.

[29]. C. Gabriel, A. Peyman, E. H. Grant, Electrical conductivity of tissue at frequencies below 1 MHz, *Physics in Medicine & Biology,* Vol. 54, 2009, pp. 4863-4878.

[30]. S. Gabriel, R. W. Lau, C. Gabriel, The dielectric properties of biological tissues: II. Measurements in the frequency range 10 Hz to 20 GHz, *Physics in Medicine & Biology,* Vol. 41, 1996, 2251.

[31]. K. Heileman, J. Daoud, M. Tabrizian, Dielectric spectroscopy as a viable biosensing tool for cell and tissue characterization and analysis, *Biosensors & Bioelectronics,* Vol. 68, 2013, pp. 348-359.

[32]. Y. Hayashi, I. Oshige, Y. Katsumoto, S. Omori, A. Yasuda, K. Asami, Dielectric inspection of erythrocyte morphology, *Physics in Medicine & Biology,* Vol. 53, 2008, 2553.

[33]. J. Q. Jaeger, R. L. Mehta, Assessment of dry weight in hemodialysis: An overview, *Journal of the American Society of Nephrology,* Vol. 10, 1999, pp. 392-403.

[34]. U. M. Moissl, P. Wabel, P. W. Chamney, I. Bosaeus, N. W. Levin, A. Bosy-Westphal, O. Korth, M. J. Muller, L Ellegård, V. Malmros, C. Kaitwatcharachai, M. K. Kuhlmann, F. Zhu, N. J. Fuller, Body fluid volume determination via body composition spectroscopy in health and disease, *Physiological Measurements,* Vol. 27, 2006, pp. 921-933.

[35]. G. Sergi, M. Bussolotto, P. Perini, I. Calliari, V. Giantin, A. Ceccon, F. Scanferla, M. Bressan, G. Moschini, G. Enzi, Accuracy of bioelectrical bioimpedance analysis for the assessment of extracellular space in healthy subjects and in fluid retention states, *Annals of Nutrition & Metabolism,* Vol. 38, 1994, pp. 158-165.

[36]. P. Deurenberg, K. van der Koy, R. Leenen, J. A. Westrate, J. C. Seidell, Sex and age specific prediction formulas for estimating body composition

from bioelectric impedance: a cross validation study, *International Journal of Obesity*, Vol. 15, 1991, pp. 17-25.

[37]. F. Seoane, S. Abtahi, F. Abtahi, L. Ellegård, G. Johannsson, I. Bosaeus, L. C. Ward, Mean expected error in prediction of total body water. A true accuracy comparison between bioimpedance spectroscopy and single frequency regression equations, *BioMed Research International*, Vol. 5, 2015, 656323.

[38]. A. Piccoli, Estimation of fluid volumes in hemodialysis patients: Comparing bioimpedance with isotopic and dilution methods, *Kidney International*, Vol. 85, 2014, pp. 738-741.

[39]. H. J. Rodriguez, R. Domenici, A. Diroll, I. Goykhman, Assessment of dry weight by monitoring changes in blood volume during hemodialysis using Crit-Line, *Kidney International*, Vol. 68, 2005, pp. 854-861.

[40]. J. G. Raimann, F. Zhu, J. Wang, S. Thijssen, M. K. Kuhlmann, P. Kotanko, N. W. Levin, G. A. Kaysen, Comparison of fluid volume estimates in chronic hemodialysis patients by bioimpedance, direct isotropic, and dilution methods, *Kidney International*, Vol. 85, 2014, pp. 898-908.

[41]. S. Castellano, I. Palomares, M. Molina, R. Pérez-García, P. Aljama, R. Ramos, J. I. Merello, Grupo ORD (Optimizando Resultados de Diálisis), Clinical, analytical and bioimpedance characteristics of persistently overhydrated haemodialysis patients, *Nefrologia*, Vol. 34, 2014, pp. 716-723.

[42]. L. D. Montgomery, W. A. Gerth, R. W. Montgomery, S. Q. Lew, M. M. Klein, J. M. Stewart, M. T. Velasquez, Monitoring intracellular, intercellular, and intravascular volume changes during fluid management procedures, *Medical & Biological Engineering & Computing*, Vol. 51, 2013, pp. 1167-1175.

[43]. J. Raimann, L. Liu, S. Tyagi, N. W. Levin, P. Kotanko, A fresh look at dry weight, *Hemodialysis International*, Vol. 12, 2008, pp. 395-405.

[44]. F. Zhu, P. Kotanko, G. J. Handelman, J. G. Raimann, L. Liu, M. Carter, M. K. Kuhlmann, E. Seibert, E. F. Leonard, N. W. Levin, Estimation of normal hydration in dialysis patients using whole body and calf bioimpedance analysis, *Physiological Measurements*, Vol. 32, 2011, pp. 887-902.

[45]. H. Jeong, C-W. Lim, H.-M. Choi, D.-J. Oh, The source of net ultrafiltration during hemodialysis is mostly the extracellular space regardless of hydration status, *Hemodialysis International*, Vol. 20, 2016, pp. 123-133.

[46]. J. Cridlig, M. Nadi, M. Kessler, Bioimpedance measurement in the kidney disease patient, in *Technical Problems in Patients on Hemodialysis* (M. G. Penido, Ed.), *InTech*, 2011.

[47]. P. W. Chamney, P. Wabel, U. M. Moissl, M. J. Müller, A. Bosy-Westphal, O. Korth, N. J. Fuller, A whole-body model to distinguish excess fluid from the hydration of major body tissues. *American Journal of Clinical Nutrition*, Vol. 85, 2007, pp. 80-89.

[48]. B. Charra, 'Dry weight in dialysis': The history of a concept, *Nephrology Dialysis Transplantation*, Vol. 7, 1998, pp. 1882-1885.

[49]. P. Wabel, U. Moissl, P. Chamney, T. Jirka, P. Machek, P. Ponce, V. Wizemann, Towards improved cardiovascular management: The necessity of combining blood pressure and fluid overload, *Nephrology Dialysis Transplantation*, Vol. 23, 2008, pp. 2965-2971.

[50]. W. J. Hannan, S. J. Cowen, K. C. Fearon, C. E. Plester, J. S. Falconer, R. A. Richardson, Evaluation of multi-frequency bio-impedance analysis for the assessment of extracellular and total body water in surgical patients, *Clinical Science*, Vol. 86, 1994, pp. 479-485.

[51]. D. P. Kotler, S. Burastero, J. Wang, Jr. R. N. Pirson, Prediction of body cell mass, fat-free mass, and total body water with bioelectrical impedance analysis: effects of race, sex and disease, *American Journal of Clinical Nutrition*, Vol. 64, 1996, pp. 4895-4975.

[52]. D. A. Schoeller, W. Dicta, E. van Santen, P. D. Klein, Validation of saliva sampling for total body water determination by H_21_5O dilution, *American Journal of Clinical Nutrition*, Vol. 35, 1982, pp. 591-594.

[53]. C. Pichonnaz, J.-P. Bassin, D. Currat, E. Martin, B. M. Jolles, Bioimpedance for oedema evaluation after total knee arthroplasty, *Physiotherapy Research International*, Vol. 18, 2013, pp. 140-147.

[54]. C. Pichonnaz, J.-P. Bassin, E. Lécureux, D. Currat, B. M. Jolles, Bioimpedance spectroscopy for swelling evaluation following total knee arthroplasty: A validation study, *BMC Musculoskeletal Disorders*, Vol. 16, 2015, pp. 101-108.

[55]. I. Beberashvili, I. Sinuani, G. Shapiro, J. Sandbank, A. Azar, H. Kadoshi, L. Feldman, Z. Averbukh, Longitudinal changes in bioimpedance phase angle reflect changes in serum IL-6 levels in maintenance hemodialysis patients, *Nutrition*, Vol. 30, 2014, pp. 297-304.

[56]. L. C. Ward, S. Czerniec, S. L. Kilbreath, Quantitative bioimpedance spectroscopy for the assessment of lymphedema, *Breast Cancer Research and Treatment*, Vol. 117, 2009, pp. 541-547.

[57]. L. C. Ward, Bioelectrical impedance analysis: proven utility in lymphedema risk assessment and therapeutic monitoring, *Lymphatic Research and Biology*, Vol. 4, 2006, pp. 51-56.

[58]. L. C. Ward, S. Czerniec, S. L. Kilbreath, Operational equivalence of bioimpedance indices and perometry for the assessment of unilateral arm lymphedema, *Lymphatic Research and Biology*, Vol. 7, 2009, pp. 81-85.

[59]. R. Gudivaka, D. A. Schoeller, R. F. Kushner, M. J. G. Bolt, Single- and multifrequency- models for bioelectrical impedance analysis of body water compartments, *Journal of Applied Physiology*, Vol. 87, 1999, pp. 1087-1096.

[60]. K. Asami, Design of a measurement cell for low-frequency dielectric spectroscopy of biological cell suspensions, *Measurement Science and Technology*, Vol. 22, 2011, 085801.

[61]. H. Morgan, T. Sun, D. Holmes, S. Gawad, N. G. Green, Single cell dielectric spectroscopy, *Journal of Physics D: Applied Physics*, Vol. 40, 2007, pp. 61-70.

[62]. T. Sun, N. G. Green, H. Morgan, Analytic and numerical modeling methods for impedance analysis of single cells on-chip, *NANO: Brief Reports and Reviews*, Vol. 3, 2008, pp. 55-63.

[63]. Y. Xu, X. Xie, Y. Duan, L. Wang, Z. Cheng, J. Cheng, A review of impedance measurements of whole cells, *Biosensors & Bioelectronics* Vol. 77, 2016, pp. 824-836.

[64]. ImpediMed SFB7: Improved Accuracy and Precision Using BIS, https://www.impedimed.com/products/sfb7-for-body-composition/

[65]. K. S. Cole, R. H. Cole, Dispersion and absorption in dielectrics, *Journal of Chemical Physics*, Vol. 9, 1941, pp. 341-351.

[66]. H. Fricke, The Maxwell-Wagner dispersion in a suspension of ellipsoids, *Journal of Physical Chemistry*, Vol. 57, 1953, pp. 934-937.

[67]. H. Looyenga, Dielectric constants of heterogeneous mixture, *Physica*, Vol. 31, 1965, pp. 401-406.

[68]. K. Asami, T. Hanai, N. Koizumi, Dielectric approach to suspensions of ellipsoidal particles covered with a shell in particular reference to biological cells, *Japanese Journal of Applied Physics*, Vol. 19, 1980, pp. 359-365.

[69]. G. A. Kaysen, F. Zhu, S. Sarkar, S. B. Heymsfield, J. Wong, C. Kait, C. Kaitwatcharachai, M. K. Kuhlmann, N. W. Levin, Estimation of total-body and limb muscle mass in hemodialysis patients by using multifrequency bioimpedance spectroscopy, *American Journal of Clinical Nutrition*, Vol. 82, 2005, pp. 988-995.

[70]. A. S. Tucker, E. A. Ross, J. Paugh-Miller, R. J. Sadleir, In vivo quantification of accumulating abdominal fluid using an electrical impedance tomography hemiarray, *Physiological Measurement*, Vol. 32, 2011, pp. 151-165.

[71]. Y. Dou, L. Liu, X. Cheng, L. Cao, L. Zuo, Comparison of bioimpedance methods for estimating total body water and intracellular water changes during hemodialysis, *Nephrology Dialysis Transplantation*, Vol. 26, 2011, pp. 3319-3324.

[72]. P. W. Chamney, P. Wabel, U. M. Moissl, M. J. Müller, A. Bosy-Westphal, O. Korth, N. J. Fuller, A whole-body model to distinguish excess fluid from the hydration of major body tissues, *American Journal of Clinical Nutrition*, Vol. 85, 2007, pp. 80-89.

[73]. E. Varlet-Marie, J. F. Brun. Clinical prediction of RBC aggregability and deformability by whole body bioimpedance measurements analyzed according to Hanai's mixture conductivity theory, *Clinical Hemorheology and Microcirculation*, Vol. 47, 2011, pp. 151-161.

[74]. E. Varlet-Marie, I. Aloulou, J. Mercier, J. F. Brun, Prediction of hematocrit and red cell deformability with whole body biological impedance, *Clinical Hemorheology and Microcirculation*, Vol. 44, 2010, pp. 237-244.

[75]. The BCM – Body Composition Monitor, Fresenius Medical Care, http://www.bcm-fresenius.com/index.html

[76]. S. Y. Semenov, G. I. Simonova, A. N. Starostin, M. G. Taran, A. E. Souvorov, A. E. Bulyshev, R. H. Svenson, A. G. Nazarov, Y. E. Sizov, V. G. Posukh, A. Pavlovsky, G. P. Tatsis, Dielectric model of cellular structures in radio frequency and microwave spectrum. Electrically interacting versus noninterecting cells, *Annals of Biomedical Engineering*, Vol. 29, 2001, pp. 427-435.

[77]. S. Y. Semenov, R. H. Svensson, G. P. Tatsis, Microwave spectroscopy of myocardial ischemia and infarction. 1. Experimental study, *Annals of Biomedical Engineering*, Vol. 28, 2000, pp. 48-54.

[78]. S. Y. Semenov, R. H. Svenson, A. E. Bulyshev, A. E. Souvorov, A. G. Nazarov, Y. E. Sizov, V. G. Posukh, A. Pavlovsky, G. P. Tatsis, Microwave spectroscopy of myocardial ischemia and infarction. 2. Biophysical reconstruction, *Annals of Biomedical Engineering*, Vol. 28, 2000, pp. 55-60.

[79]. G. Sauerbrey, Verwendung von Schwingquarzen zur Wägung dünner Schichten und zur Mikrowägung, *Zeitschrift für Physik*, Vol. 155, 1959, pp. 206-222.

[80]. K. Kanazawa, J. G. Gordon II, Frequency of a quartz microbalance in contact with liquid, *Analytical Chemistry*, Vol. 57, 1985, pp. 1770-1771.

[81]. S. J. Martin, R. W. Cernosek, J. J. Spates, Sensing liquid properties with shear-mode resonator sensors, in *Proceedings of the International Conference On Solid-State Sensors And Actuators And Eurosensors IX (Transducers'95)*, Stockholm, Sweden, 1995, pp. 712-715.

[82]. S. Martin, V. E. Granstaff, G. C. Frye, Characterization of a quartz crystal microbalance with simultaneous mass and liquid loading, *Analytical Chemistry*, Vol. 63, 1991, pp. 2272-2281.

[83]. European Network on New Sensing Technologies for Air-Pollution Control and Environmental Sustainability EuNetAir, http://www.cost.eunetair.it

[84]. The Quality of Air, Comprehensive Analytical Chemistry (M. de la Guardia, S. Armenta, Eds.), Vol. 73, *Elsevier*, 2016.

[85]. Real-time MOUDI Quartz Crystal Microbalance (QCM) Impactor http://www.tsi.com/msp_cascade_impactors/

[86]. H. Muramatsu, E. Tamiya, I. Karube, Computation of equivalent circuit parameters of quartz crystals in contact with liquids and study of liquid properties, *Analytical Chemistry*, Vol. 60, 1988, pp. 2142-2146.

[87]. A. P. M. Glassford, Response of a quartz crystal microbalance to a liquid deposit, *Journal of Vacuum Science & Technology*, Vol. 15, 1978, 1836.

[88]. M. Rodahl, F. Höök, A. Krozer, P. Brzezinski, B. Kasemo, Quartz crystal microbalance setup for frequency and Q-factor measurements in gaseous and liquid environments, *Review of Scientific Instruments*, Vol. 66, 1995, pp. 3924-3930.

[89]. SRS, http://www.thinkSRS.com

[90].   R. E. Speight, M. A. Cooper, A survey of the 2010 quartz crystal microbalance literature, *Journal of Molecular Recognition*, Vol. 25, 2012, pp. 451-473.

[91].   D. Johannsmann, Quartz Crystal Microbalance in Soft Matter Research. Fundamentals and Modeling, *Springer*, 2015.

[92].   Rodahl, F. Höök, A. Krozer, P. Brzezinski, B. Kasemo, Quartz crystal microbalance setup for frequency and Q-factor measurements in gaseous and liquid environments, *Review of Scientific Instruments*, Vol. 66, 1995, pp. 3924-3930.

[93].   M. Rodahl, B. Kasemo, A simple setup to simultaneously measure the resonant frequency and the absolute dissipation factor of a quartz crystal microbalance, *Review of Scientific Instruments*, Vol. 67, 1996, pp. 3924-3930.

[94].   M. Rodahl, F. Höök, B. Kasemo, QCM operation in liquids: an explanation of measured variations in frequency and Q factor with liquid conductivity, *Analytical Chemistry*, Vol. 68, 1996, pp. 2219-2227.

[95].   M. Rodahl, B. Kasemo, On measurement of thin overlayers with the quartz crystal microbalance, *Sensors and Actuators A: Physical*, Vol. 54, 1996, pp. 448-456.

[96].   QSense, Biolin Scientific, https://www.biolinscientific.com/qsense

[97].   D. Johannsmann, Viscoelastic, mechanical, and dielectric measurements on complex samples with the quartz crystal microbalance, *Physical Chemistry Chemical Physics*, Vol. 10, 2008, pp. 4516-4534.

[98].   D. Johannsmann, T. Mathauer, G. Wegner, W. Knoll, Viscoelastic properties of thin films probed with a quartz-crystal resonator, *Physical Review B*, Vol. 46, 1992, pp. 7808-7815.

[99].   C. E. Reed, K. K. Kanazawa, J. H. Kaufman, Physical description of a viscoelastically loaded AT-cut quartz resonator, *Journal of Applied Physics*, Vol. 68, 1990, 1993.

[100]. M. Rodahl, F. Höök, C. Fredriksson, C. Keller, A. Krozer, P. Brzezinski, M. Voinova, B. Kasemo, Simultaneous frequency and dissipation factor QCM measurements of biomolecular adsorption and cell adhesion, *Faraday Discussions*, Vol. 107, 1996, pp. 229-246.

[101]. R. Lucklum, P. Hauptmann, The quartz crystal microbalance: Mass sensitivity, viscoelasticity and acoustic amplification, *Sensors and Actuators B: Chemical*, Vol. 70, 2000, pp. 30-36.

[102]. R. Lucklum, P. Hauptmann, Acoustic microsensors-the challenge behind microgravimetry, *Analytical and Bioanalytical Chemistry*, Vol. 384, 2006, pp. 667-682.

[103]. A. Janshoff, C. Steinem, Quartz crystal microbalance for bioanalytical applications, in Sensors Update (H. Baltes, J. Hesse, J. G. Korvink, Eds.), Vol. 9, *Wiley-VCH*, 2001.

[104]. J. Wegener, A. Janshoff, C. Steinem, The quartz crystal microbalance as a novel means to study cell-substrate interactions in situ, *Cell Biochemistry and Biophysics*, Vol. 34, 2001, pp. 121-151.

[105]. K. Kanazawa, C. W. Frank, J. Hardesty, Resonances of soft films under liquids on the QCM, *ECS Transactions*, Vol. 16, 2008, pp. 419-429.

[106]. K. A. Marx, Quartz crystal microbalance: a useful tool for studying thin polymer films and complex biomolecular systems at the solution-surface interface, *Biomacromolecules*, Vol. 4, 2003, pp. 1099-1120.

[107]. D. Johannsmann, Viscoelastic analysis of organic thin films on quartz resonators, *Macromolecular Chemistry and Physics*, Vol. 200, 1999, pp. 501-516.

[108]. H. L. Bandey, S. J. Martin, R. W. Cernosek, A. R. Hillman, Modeling the responses of thickness-shear mode resonators under various loading conditions, *Analytical Chemistry*, Vol. 71, 1999, pp. 2205-2214.

[109]. M. V. Voinova, M. Rodahl, M. Jonson, B. Kasemo, Viscoelastic acoustic response of layered polymer films at fluid-solid interfaces: Continuum mechanics approach, *Physica Scripta*, Vol. 759, 1999, pp. 391-396.

[110]. M. V. Voinova, M. Jonson, B. Kasemo, "Missing mass" effect in biosensors QCM applications, *Biosensors & Bioelectronics*, Vol. 17, 2002, pp. 835-841.

[111]. R. Lucklum, P. Hauptmann, Determination of polymer shear modulus with quartz crystal resonators, *Faraday Discussions*, Vol. 107, 1997, pp. 123-130.

[112]. A. P. Borovikov, Measurement of viscosity of media by means of shear vibration of plane piezoresonators, *Instruments and Experimental Technics*, Vol. 19, 1976, pp. 223-224.

[113]. M. V. Voinova, M. Jonson, B. Kasemo, Dynamics of viscous amphiphilic films supported by elastic solid substrates, *Journal of Physics C*, Vol. 9, 1997, pp. 7799-7808.

[114]. M. V. Voinova, Modelling of the response of acoustic piezoelectric resonators in biosensor applications – Part 1: The general theoretical analysis, *Journal of Sensors and Sensor Systems*, Vol. 4, 2015, pp. 137-142.

[115]. A. Vikström, M. V. Voinova, Soft-film dynamics of SH-SAW sensors in viscous and viscoelastic fluids, *Sensing and Bio-Sensing Research*, Vol. 11, 2016, pp. 78-85.

[116]. D. R. Thévenot, K. Toth, R. A. Durst, G. S. Wilson, Electrochemical biosensors: recommended definitions and classification, *Biosensors & Bioelectronics*, Vol. 16, 2001, pp. 121-131.

[117]. E. Sackmann, Supported membranes: Scientific and practical applications, *Science*, Vol. 271, 1996, pp. 43-48.

[118]. A. N. Parikh, J. T. Groves, Materials science of supported lipid membranes, *MRS Bulletin*, Vol. 31, 2006, pp. 507-512.

[119]. M. Tanaka, Polymer-supported membranes: Physical models of cell surfaces, *MRS Bulletin*, Vol. 31, 2006, pp. 513-520.

[120]. A. Janshoff, H.-J. Galla, C. Steinem, Biochemical applications of solid supported membranes on gold surfaces: quartz crystal microbalance and impedance analysis, in Planar Lipid Bilayers (BLMs) and Their Application (H. T. Tien, A. Ottova, Eds.), *Elsevier*, 2003, pp. 991-1016.

[121]. C. A. Keller, B. Kasemo, Surface specific kinetics of lipid vesicle adsorption measured with a quartz crystal microbalance, *Biophysical Journal*, Vol. 75, 1998, pp. 1397-1402.

[122]. J. Nissen, K. Jacobs, J. O. Rädler, Interface dynamics of lipid membrane spreading on solid surfaces, *Physical Review Letters*, Vol. 86, 2001, pp. 1904-1907.

[123]. M. Stelzle, G. Weissmüller, E. Sackmann, On the application of supported bilayers as receptive layers for biosensors with electrical detection, *Journal of Physical Chemistry*, Vol. 97, 1993, pp. 2974-2981.

[124]. C. Steinem, A. Janshoff, W.-P. Ulrich, M. Sieber, H.-J. Galla, Impedance analysis of supported lipid bilayer membranes: a scrutiny of different preparation techniques, *Biochimica et Biophysica Acta*, Vol. 1279, 1996, pp. 169-180.

[125]. A.-S. Cans, F. Höök, O. Shupliakov, A. G. Ewing, P. S. Eriksson, L. Brodin, O. Orwar, Measurement of the dynamics of exocytosis and vesicle retrieval at cell populations using a quartz crystal microbalance, *Analytical Chemistry*, Vol. 73, 2001, pp. 5805-5811.

[126]. E. Reimhult E., F. Höök, B. Kasemo, Rupture pathway of phosphatidylcholine liposomes on silicon dioxide, *International Journal of Molecular Sciences*, Vol. 10, 2009, pp. 1683-1696.

[127]. K. Glasmästar, C. Larsson, F. Hook, B. Kasemo, Protein adsorption on supported phospholipid bilayers, *Journal of Colloid and Interface Sciences*, Vol. 246, 2002, pp. 40-47.

[128]. A. S. Andersson, K. Glasmästar, D. Sutherland, U. Lidberg, B. Kasemo, Cell adhesion on supported lipid bilayers, *Journal of Biomedical Materials Research Part A*, Vol. 64A, 2003, pp. 622-629.

[129]. D. Thid, K. Holm, P. S. Eriksson, J. Ekeroth, B. Kasemo, J. Gold, Supported phospholipid bilayers as a platform for neural progenitor cell culture, *Journal of Biomedical Materials Research Part A*, Vol. 84A, 2008, pp. 940-953.

[130]. R. P. Richter, R. Berat, A. R. Brisson, Formation of solid-supported lipid bilayers: An integrated view, *Langmuir*, Vol. 22, 2006, pp. 3497-3505.

[131]. F. Hook, J. Voros, M. Rodahl, R. Kurrat, P. Boni, J. J. Ramsden, M. Textor, N. D. Spencer, P. Tengvall, J. Gold, B. Kasemo, A comparative study of protein adsorption on titanium oxide surfaces using in situ ellipsometry, optical waveguide light mode spectroscopy, and quartz crystal microbalance/dissipation, *Colloids and Surfaces B: Biointerfaces*, Vol. 24, 2002, pp. 155-170.

[132]. C. Merz, W. Knoll, M. Textor, E. Reimhult, Formation of supported bacterial lipid membrane mimics, *Biointerphases*, Vol. 3, 2008, pp. FA41-FA50.

[133]. C. Fredriksson, S. Khilman, B. Kasemo, D. M. Steel, In vitro real time characterization of cell attachment and spreading, *The Journal of Materials Science: Materials in Medicine*, Vol. 9, 1998, pp. 785-788.

[134]. N.-J. Cho, C. W. Frank, B. Kasemo, F. Höök, Quartz crystal microbalance with dissipation monitoring of supported lipid bilayers on various substrates, *Nature Protocols*, Vol. 5, 2010, pp. 1096-1106.

[135]. F. Höök, B. Kasemo, The QCM-D technique for probing biomacromolecular recognition reactions, *Springer Series on Chemical Sensors and Biosensors*, Vol. 5, 2007, pp. 425-447.

[136]. N. J. Cho, K. H. Cheong, C. Lee, C. W. Frank, J. S. Glenn, Binding dynamics of hepatitis C virus NS5A amphipathic peptide to cell and model membranes, *Journal of Virology*, Vol. 81, 2007, pp. 6682-6689.

[137]. N. J. Cho, S. J. Cho, K. H. Cheong, J. S. Glenn, C. W. Frank, Employing an amphipathic viral peptide to create a lipid bilayer on Au and TiO_2, *Journal of the American Chemical Society*, Vol. 129, 2007, pp. 10050-10051.

[138]. M. A. Cooper, V. T. A. Singleton, A survey of the 2001 to 2005 quartz crystal microbalance biosensor literature: Applications of acoustic physics to the analysis of biomolecular interactions, *Journal of Molecular Recognition*, Vol. 20, 2007, pp. 154-184.

[139]. S. X. Liu, J.-T. Kim, Application of Kevin-Voigt model in quantifying whey protein adsorption on polyethersulfone using QCM-D, *The Journal of the Association for Laboratory Automation*, 2009, pp. 213-220.

[140]. E. Li, M. Merzlyakov, J. Lin, P. Searson, K. Hristova, Utility of surface-supported bilayers in studies of transmembrane helix dimerization, *Journal of Structural Biology*, Vol. 168, 2009, pp. 53-60.

[141]. N.-J. Cho, K. K. Kanazawa, Jeffrey S. Glenn, C. W. Frank, Employing two different quartz crystal microbalance models to study changes in viscoelastic behavior upon transformation of lipid vesicles to a bilayer on a gold surface, *Analytical Chemistry*, Vol. 79, 2007, pp. 7027-7035.

[142]. M. Höpfner, R. Ulrich, B. Gerd, Biosensor-based evaluation of liposomal behavior in the target binding process, *Journal of Liposome Research*, Vol. 18, 2008, pp. 71-82.

[143]. I. Pfeffer, S. Petronis, I. Koper, B. Kasemo, M. Zach, Vesicle adsorption and phospholipid bilayer formation on topographically and chemically nanostructured surfaces, *Journal of Physical Chemistry B*, Vol. 114, 2010, pp. 4623-4631.

[144]. U. Seifert, R. Lipowsky, Adhesion of vesicles, *Physical Review A*, Vol. 42, 1990, pp. 4768-4771.

[145]. U. Seifert, Adhesion of vesicles in two dimensions, *Physical Review A*, Vol. 43, 1991, pp. 6803-6814.

[146]. E. Lüthgens, A. Herrig, K. Kastl, C. Steinem, B. Reiss, J. Wegener, B. Pignataro, A. Janshoff, Adhesion of liposomes: A quartz crystal microbalance study, *Measurement Science and Technology*, Vol. 14, 2003, pp. 1865-1875.

[147]. M. V. Voinova, On mass loading and dissipation measured with acoustic wave sensors: A review, *Journal of Sensors*, Vol. 2009, 2009, 943125.

[148]. A. Bossi, S. A. Piletsky, E. V. Piletska, P. G. Righetti, A. P. F. Turner, Surface-grafted molecularly imprinted polymers for protein recognition, *Analytical Chemistry*, Vol. 73, 2001, pp 5281-5286.

[149]. S. A. Piletsky, S. Alcock, A. P. F. Turner, Molecular imprinting at the edge of the third millennium, *Trends in Biotechnology*, Vol. 19, 2001, pp. 9-12.

[150]. O. Hayden, C. Haderspöck, S. Krassnig, X. Chen, F. L. Dickert, Surface imprinting strategies for the detection of trypsin, *Analyst*, Vol. 131, 2006, pp. 1044-1050.

[151]. M. Hussain, J. Wackerlig, P. A. Lieberzeit, Biomimetic strategies for sensing biological species, *Biosensors*, Vol. 3, 2013, pp. 89-107.

[152]. T.-Y. Lin, C.-H. Hue, T.-C. Chou, Determination of albumin concentration by MIP-QCM sensor, *Biosensors & Bioelectronics*, Vol. 20, 2004, pp. 75-81.

[153]. W. Cai, S. Xie, J. Zhang, D. Tang, Y. Tang, An electrochemical impedance biosensor for Hg2 + detection based on DNA hydrogel by coupling with DNAzyme-assisted target recycling and hybridization chain reaction, *Biosensors & Bioelectronics*, Vol. 98, 2017, pp. 466-472.

[154]. F. Höök, A. Ray, B. Nordén, B. Kasemo, Characterization of PNA and DNA immobilization and subsequent hybridization with DNA using acoustic-shear-wave attenuation measurements, *Langmuir*, Vol. 17, 2001, pp. 8305-8312.

[155]. A. Granéli, M. Edvardsson, F. Höök, DNA-Based Formation of a Supported, Three-Dimensional Lipid Vesicle Matrix Probed by QCM-D and SPR, *A European Journal of Chemical Physics and Physical Chemistry*, Vol. 5, 2004, pp. 729-733.

[156]. S. Svedhem, D. Dahlborg, J. Ekeroth, J. Kelly, F. Höök, J. Gold, In situ peptide-modified supported lipid bilayers for controlled cell attachment, *Langmuir*, Vol. 19, 2003, pp. 6730-6736.

[157]. S. Seker, A. E. Elcin, Y. M. Elcin, Real-time monitoring of mesenchymal stem cell responses to biomaterial surfaces and to a model drug by using quartz crystal microbalance, *Artificial Cells, Nanomedicine, and Biotechnology*, Vol. 44, 2016, pp. 1722-1732.

[158]. N. Tymchenko, E. Nilebäck, M. V. Voinova, J. Gold, B. Kasemo, S. Svedhem, Reversible changes in cell morphology due to cytoskeletal rearrangements measured in real-time by QCM-D, *Biointerphases*, Vol. 43, 2012, pp. 1-4.

[159]. T. W. Chung, Y. C. Tyan, R. H. Lee, C. W. Ho, Determining early adhesion of cells on polysaccharides/PCL surfaces by a quartz crystal microbalance, *The Journal of Materials Science: Materials in Medicine*, Vol. 23, 2012, pp. 3067-3073.

[160]. J. Wegener, A. Janshoff, C. Steinem, The quartz crystal microbalance as a novel means to study cell-substrate interactions in situ, *Cell Biochemistry and Biophysics*, Vol. 34, 2001, pp. 121-151.

[161]. T. Zhou, K. A. Marx, A. H. Dewilde, D. McIntosh, S. J. Braunhut, Dynamic cell adhesion and viscoelastic signatures distinguish normal from malignant human mammary cells using quartz crystal microbalance, *Analytical Biochemistry*, Vol. 421, 2012, pp. 164-171.

[162]. J. Y. Chen, L. S. Penn, J. Xi, Quartz crystal microbalance: Sensing cell-substrate adhesion and beyond, *Biosensors & Bioelectronics*, Vol. 99, 2018, pp. 593-602.

[163]. N. Z. Piety, W. H. Reinhart, P. H. Pourreau, R. Abidi, S. S. Shevkoplyas, Shape matters: The effect of red blood cell shape on perfusion of an artificial microvascular network, *Transfusion*, Vol. 56, 2015, pp. 844-851.

[164]. R. M. Albu, Surface properties and compatibility with blood of new quaternized polysulfones, *Journal of Biomaterials and Nanobiotechnology*, Vol. 2, 2011, pp. 114-123.

[165]. Bernhardt, L. Ivanova, P. Langehanenberg, B. Kemper, G. von Bally, Application of digital holographic microscopy to investigate the sedimentation of intact red blood cells and their interaction with artificial surfaces, *Bioelectrochemistry*, Vol. 73, 2008, pp. 92-96.

[166]. C. Zandén, M. V. Voinova, J. Gold, D. Mörsdorf, I. Bernhardt, J. Liu, Surface characterisation of oxygen plasma treated electrospun polyurethane fibres and their interaction with red blood cells, *European Polymer Journal*, Vol. 48, 2012, pp. 472-482.

[167]. R. Grzhibovskis, E. Krämer, I. Bernhardt, B. Kemper, C. Zanden, N. V. Repin, B. V. Tkachuk, M. V. Voinova, Shape of red blood cells in contact with artificial surfaces, *European Biophysics Journal*, Vol. 46, 2017, pp. 141-148.

[168]. Medical Coatings and Deposition Technologies (D. Glocker, S. Ranade, Eds.), *Scrivener Publishing, Wiley*, 2016.

[169]. M. Andersson, J. Andersson, A. Sellborn, M. Berglin, B. Nilsson, H. Elwing, Quartz crystal microbalance-with dissipation monitoring (QCM-D) for real time measurements of blood coagulation density and immune complement activation on artificial surfaces, *Biosensors & Bioelectronics*, Vol. 21, 2005, pp. 79-86.

[170]. K. S. Lavery, C. Rhodes, A. Mcgraw, M. J. Eppihimer, Anti-thrombotic technologies for medical devices, *Advanced Drug Delivery Reviews*, Vol. 112, 2017, pp. 2-11.

[171]. R. Biran, D. Pond, Heparin coating for improving blood compatibility of medical devices, *Advanced Drug Delivery Reviews*, Vol. 112, 2017, pp. 12-23.

[172]. K. Christensen, R. Larsson, H. Emanuelsson, G. Elgue, A. Larsson, Heparin coating of the stent graft-effects on platelets, coagulation and complement activation, *Biomaterials*, Vol. 22, 2001, pp. 349-355.

[173]. S. Sagedal, B. J. Witczak, K. Osnes, A. Hartmann, I. Os, L. Eikvar, O. Klingenberg, F. Brosstad, A heparin-coated dialysis filter (AN69 ST) does not reduce clotting during hemodialysis when compared to a

conventional polysulfone filter (F×8), *Blood Purification*, Vol. 32, 2011,
pp. 151-155.

[174]. F. Ziai, T. Benesch, K. Kodras, I. Neumann, L. Dimopoulos-Xicki,
M. Haas, The effect of oral anticoagulation on clotting during
hemodialysis, *Kidney International*, Vol. 68, 2005, pp. 862-866.

[175]. M. Andersson, A. Sellborn, C. Fant, C. Gretzer, H. Elwing, Acoustics of
blood plasma on solid surfaces, *Journal of Biomaterials Science,
Polymer Edition*, Vol. 13, 2002, pp. 907-917.

[176]. Y. Yang, W. Zhang, Z. Guo, Z. Zhang, H. Zhu, R. Yan, L. Zhou, Stability
enhanced, repeatability improved Parylene-C passivated on QCM sensor
for aPTT measurement, *Biosensors & Bioelectronics*, Vol. 98, 2017,
pp. 41-46.

[177]. Acute Blood Purification (H. Suzuki, H. Hirasawa, Eds.), *Scitech Book
News*, 2010.

[178]. R. D. Frank, J. Weber, H. Drebach, H. Thelen, C. Weiss, J. Floege, Role
of contact system activation in hemodialyzer-induced thrombogenicity,
Kidney International, Vol. 60, 2001, pp. 1972-1981.

[179]. R. S. Lakshmanan, V. Efremov, S. Cullen, B. Byrne, A. J. Killard,
Monitoring the effects of fibrinogen concentration on blood coagulation
using quartz crystal microbalance (QCM) and its comparison with
thromboelastography, *Proceedings of SPIE*, Vol. 8765, 2013, 87650Q.

[180]. V. Efremov, R. S. Lakshmanan, B. Byrne, S. M. Cullen, A. J. Killard,
Modelling of blood component flexibility using quartz crystal
microbalance, *Journal of Biorheology*, Vol. 28, 2014, pp. 45-54.

[181]. V. Efremov, A. J. Killard, B. Byrne, R. S. Lakshmanan, The modelling
of blood coagulation using the quartz crystal microbalance, *Journal of
Biomechanics*, Vol. 46, 2013, pp. 437-442.

[182]. V. Bowbrick, D. P. Mikhailidis, G. Stansby, The use of citrated whole
blood in thromboelastography, *Anesthesia & Analgesia*, Vol. 90, 2000,
pp. 1086-1088.

[183]. V. Nielsen, B. Cohen, E. Cohen, Effects of coagulation factor deficiency
on plasma coagulation kinetics determined via thromboelastography
critical roles of fibrinogen and factors II, VII, X and XII, *Acta
Anaesthesiologica Scandinavica*, Vol. 49, 2005, pp. 222-231.

[184]. M. Thakur, A. B. Ahmed, A review of thromboelastography,
*International Journal of Perioperative Ultrasound and Applied
Technologies*, Vol. 1, 2012, pp. 25-29.

[185]. M. Hussain, S. Sinn, M. Zeilinger, H. Northoff, P. A. Lieberzeit,
F. K. Gehring, Blood coagulation thromboplastine time measurements
on a nanoparticle coated quartz crystal microbalance biosensor in
excellent agreement with standard clinical methods, *Biosensors &
Bioelectronics*, Vol. 4, 2013, 139.

[186]. S. Oberfrank, H. Drechsel, S. Sinn, H. Northoff, F. K. Gehring,
Utilisation of quartz crystal microbalance sensors with dissipation

(QCM-D) for a Clauss fibrinogen assay in comparison with common coagulation reference methods, *Sensors*, Vol. 16, 2016, 282.

[187]. S. Sinn, L. Müller, H. Drechsel, M. Wandel, H. Northoff, G. Ziemer, H. P Wendel, F. K. Gehring, Platelet aggregation monitoring with a newly developed quartz crystal microbalance system as an alternative to optical platelet aggregometry, *Analyst*, Vol. 135, 2010, pp. 2930-2938.

[188]. J. Lutz, J. Menke, D. Sollinger, H. Schinzel, K. Thürmel, Haemostasis in chronic kidney disease, *Nephrology Dialysis Transplant*, Vol. 29, 2014, pp. 29-40.

[189]. L. Müller, S. Sinn, H. Drechsel, C. Ziegler, H. P. Wendel, H. Northoff, F. K. Gehring, Investigation of Prothrombin Time in human whole-blood samples with a quartz crystal biosensor, *Analytical Chemistry*, Vol. 82, 2010, pp. 658-663.

[190]. N. Wisniewski, F. Moussy, W. M. Reichert, Characterization of implantable biosensor membrane biofouling, *Journal of Analytical Chemistry*, Vol. 366, 2000, pp. 611-621.

[191]. A. Hirano, A. Hirano, S. Kida, K. Yamamoto, K. Sakai, Experimental evaluation of flow and dialysis performance of hollow-fiber dialyzers with different packing densities, *Journal of Artificial Organs*, Vol. 15, 2011, pp. 168-175.

[192]. M. Hiwatari, K. Yamamoto, M. Hayama, F. Kohori, K. Sakai, Evaluation of local membrane fouling in hemodialyzer, *ASAIO Journal*, Vol. 50, 2004, p. 177.

[193]. M. Hayama, K.-I. Yamamoto, F. Kohori, K. Sakai, How polysulfone dialysis membranes containing polyvinylpyrrolidone achieve excellent biocompatibility?, *Journal of Membrane Science*, Vol. 234, 2004, pp. 41-49.

[194]. F. Baohong, C. Cheng, L. Li, J. Cheng, W. Zhao, C. Zhao, Surface modification of polyethersulfone membrane by grafting bovine serum albumin, *Fibers and Polymers*, Vol. 11, 2010, pp. 960-966.

[195]. B. Xie, R. Zhang, H. Zhang, A. Xu, Y. Deng, Yalin Lv, F. Deng, S. Wei, Decoration of heparin and bovine serum albumin on polysulfone membrane assisted via polydopamine strategy for hemodialysis, *Journal of Biomaterials Science Polymer Edition*, Vol. 27, 2016, pp. 880-897.

[196]. X. J. Huang, D. Guduru, Z. K. Xu, J. Veinken, T. Groth, Blood compatibility and permeability of heparin-modified polysulfone as potential membrane for simultaneous hemodialysis and LDL removal, *Macromolecular Bioscience*, Vol. 11, 2011, pp. 131-140.

[197]. I. Kazuhiko, F. Kikuko, I. Yasuhiko, N. Nakabayashi, Modification of polysulfone with phospholipid polymer for improvement of the blood compatibility. Part 1. Surface characterization, *Biomaterials*, Vol. 20, 1999, pp. 1545-1551.

[198]. J.-H. Jiang, L.-P. Zhu, X.-L. Li, Y.-Y. Xu, B.-K. Zhu, Surface modification of PE porous membranes based on the strong adhesion of polydopamine and covalent immobilization of heparin, *Journal of Membrane Science*, Vol. 364, 2010, pp. 194-202.

[199]. A. Sweity, Y. Wang, M. S. Ali-Shtayeh, F. Yang, A. Bick, G. Oron, M. Herzberg, Relation between EPS adherence, viscoelastic properties, and MBR operation: Biofouling study with QCM-D, *Water Research*, Vol. 45, 2011, pp. 6430-6440.

[200]. W. Zhao, J. Huang, B. Fang, S. Nie, Nan Yi, B. Su, L. Haifeng, Z. Changsheng, Modification of polyethersulfone membrane by blending semi-interpenetrating network polymeric nanoparticles, *Journal of Membrane Science*, Vol. 369, 2011, pp. 258-266.

[201]. N. Arahman, T. Maruyma, T Sotani, H. Matsuyama, Fouling reduction of a poly(ether sulfone) hollow-fiber membrane with a hydrophilic surfactant prepared via non-solvent-induced phase separation, *Applied Polymer Science*, Vol. 111, 2009, pp. 1653-1658.

[202]. B. Su, S. Sun, C. Zhao, Polyethersulfone hollow fiber membranes for hemodialysis, Chapter 4, in Progress in Hemodialysis-From Emergent Biotechnology to Clinical Practice (A. Carpi, C. Donadio, G. Tramonti, Eds.), *Intech*, 2011, pp. 65-92.

[203]. K. Lipponen, Y. Liu, P. W. Stege, K. Öörni, P. T. Kovanen, M. L. Riekkola, Capillary electrochromatography and quartz crystal microbalance, valuable techniques in the study of heparin-lipoprotein interactions, *Analytical Biochemistry*, Vol. 424, 2012, pp. 71-78.

[204]. M. Doi, S. F. Edwards, The Theory of Polymer Dynamics, International Series of Monographs on Physics, *Clarendon Press*, 1986.

[205]. K. Länge, B. E. Rapp, M. Rapp, Surface acoustic wave biosensors: A review, *Analytical and Bioanalytical Chemistry*, Vol. 391, 2008, pp. 1509-1519.

[206]. D. S. Ballantine, R. M. White, S. J. Martin, A. J. Ricco, E. T. Zellers, G. C. Frye, H. Wohltjen, Acoustic wave sensors: Theory, design, and physico-chemical applications, in Applications of Modern Acoustics, *Academic Press*, 1997.

[207]. A. J. Ricco, S. J. Martin, Acoustic wave viscosity sensor, *Applied Physics Letters*, Vol. 50, 1987, pp. 1474-1476.

[208]. E. Gizeli, N. J. Goddard, C. W. Lowe, A Love plate biosensor utilizing a polymer layer, *Sensors & Actuators B: Chemical*, Vol. 6, 1992, pp. 131-137.

[209]. E. Gizeli, A. C. Stevenson, J. Goddard, C. R. Lowe, A novel Love-plate acoustic sensor utilizing polymer overlayers, *IEEE Transactions on Ultrasonics, Ferroelectrics, and Frequency Control*, Vol. 39, 1992, pp. 657-659.

[210]. G. Kovacs, M. J. Vellekoop, R. Haueis, G. W. Lubking, A. Venema, A love wave sensor for (bio)chemical sensing in liquids, *Sensors and Actuators A: Physical*, Vol. 43, 1994, pp. 38-43.

[211]. K. Melzak, E. Gizeli, Love wave biosensors, in Handbook of Biosensors and Biochips (C. R. Lowe, D. Cullen, H. W. Weetall, I. Karube, Eds.), *John Wiley & Sons*, 2007.

[212]. M. M. I. R. Gaso, Y. Jiménez, L. A. Francis, A. Arnau, Love wave biosensors: a review, in State of the Art in Biosensors – General Aspects (T. Rinken, Ed.), *INTEC*, 2013.

[213]. E. Gizeli, M. Liley, C. R. Lowe, H. Vogel, Detection of supported lipid bilayers with the acoustic Love waveguide device: Application to biosensors, *Sensors and Actuators B-Chemical*, Vol. 34, 1996, pp. 295-300.

[214]. E. Gizeli, F. Bender, A. Rasmusson, K. Saha, F. Josse, R. Cernosek, Sensitivity of the acoustic waveguide biosensor to protein binding as a function of the waveguide properties, *Biosensors & Bioelectronics*, Vol. 18, 2003, pp. 71399-71406.

[215]. K. Saha, F. Bender, A. Rasmusson, E. Gizeli, Probing the viscoelasticity and mass of a surface-bound protein layer with an acoustic waveguide device, *Langmuir*, Vol. 19, 2003, pp. 1304-1311.

[216]. M. Saitakis, E. Gizeli, Acoustic sensors as a biophysical tool for probing cell attachment and cell/surface interactions, *Cellular and Molecular Life Sciences*, Vol. 69, 2011, pp. 357-371.

[217]. E. Gizeli, H. Mehta, C. R. Lowe, Calibration of the Love wave sensor utilizing phospholipid bilayers, *Chem. Biol. Sensors and Analyt. Electrochem. Methods Proceedings*, Vol. 97, 1997.

[218]. G. Kovacs, A. Venema, Theoretical comparison of sensitivities of acoustic shear wave modes for (bio)chemical sensing in liquids, *Applied Physics Letters*, Vol. 61, 1992, pp. 639-659.

[219]. A. Tsortos, G. Papadakis, E. Gizeli, On the hydrodynamic nature of DNA acoustic sensing, *Analytical Chemistry*, Vol. 88, 2016, pp. 6472-6478.

[220]. G. McHale, M. I. Newton, F. Martin, K. Melzak, E. Gizeli, Resonant conditions for Love wave guiding layer thickness, *Applied Physics Letters*, Vol. 79, 2001, pp. 3542-3543.

[221]. G. McHale, M. I. Newton, F. Martin, Layer guided shear horizontally polarized acoustic plate modes, *Journal of Applied Physics*, Vol. 91, 2002, pp. 5735-5744.

[222]. M. I. Newton, F. Martin, K. Melzak, E. Gizeli, G. McHale, Harmonic Love devices for biosensing applications, *Electronics Letters*, Vol. 37, 2001, pp. 340-341.

[223]. K. Kalantar-Zadeh, W. Wlodarski, Y. Y. Chen, B. N. Fry, K. Galatsis, Novel Love mode surface acoustic wave based immunosensors, *Sensors & Actuators B-Chemical*, Vol. 91, 2003, pp. 143-147.

[224]. B. Rapp, K. Länge, M. Rapp, A. Guber, Surface acoustic wave biosensors for biomedical applications, in *Proceedings of the Jahrestagung der Deutschen Gesellschaft für Biomedizinische Technik (DGBMT'07)*, Aachen, Germany, 26-29 September 2007.

[225]. J. Freudenberg, M. von Schickfus, S. Hunklinger, A SAW immunosensor for operation in liquid using a SiO2 protective layer, *Sensors & Actuators B: Chemical*, Vol. 76, 2001, pp. 147-151.

[226]. A. Kordas, G. Papadakis, D. Milioni, J. Champ, S. Descroix, E. Gizeli, Rapid Salmonella detection using an acoustic wave device combined

with the RCA isothermal DNA amplification method, *Sensing and Bio-Sensing Research,* Vol. 11, 2016, pp. 121-127.

[227]. M. Gianneli, K. Tsougeni, A. Grammoustianou, A. Tserepi, E. Gogolides, E. Gizeli, Nanostructured PMMA-coated Love wave device as a platform for protein adsorption studies, *Sensors and Actuators B: Chemical,* Vol. 236, 2016, pp. 583-590.

[228]. L. D. Landau, E. M. Lifshitz, Fluid Mechanics, Course of Theoretical Physics, 2nd Ed., Vol. 6, *Pergamon Press,* 1987.

[229]. M. Oberleiter, QCM-ECIS: Combined viscoelastic and dielectric sensing of cells, in Label-Free and Multi-Parametric Monitoring of Cell-Based Assays with Substrate-Embedded Sensors, Springer Theses, *Springer,* 2018.

[230]. A. Hinz, H. J. Galla, Impedance Spectroscopy and Quartz Crystal Microbalance, in Molecular Sensors for Cardiovascular Homeostasis (D. H. Wang, Ed.), *Springer,* Boston, MA, 2007.

[231]. A. Janshoff, J. Wegener, M. Sieber, H. J. Galla, Double-mode impedance analysis of epithelial cell monolayers cultured on shear wave resonators, *European Biophysical Journal,* Vol. 25, 1996, pp. 93-103.

[232]. I. Giaever I, C. R. Keese, Micromotion of mammalian cells measured electrically, *Proceedings of the National Academy of Sciences of USA,* Vol. 88, 1991, pp. 7896-7900.

[233]. Q. Liu, C. Wu, N. Hu, J. Zhou, P. Wang, Cell-based biosensors and their application in biomedicine, *Chemical Reviews,* Vol. 114, 2014, pp. 6423-6461.

[234]. L. Wang, H. Wang, L. Wang, K. Mitchelson, Z. Yu, J. Cheng, Analysis of the sensitivity and frequency characteristics of coplanar electrical cell-substrate impedance sensors, *Biosensors & Bioelectronics,* Vol. 24, 2008, pp. 14-21.

[235]. Q. Xie, C. Xiang, X. Yang, Y. Zhang, M. Li, S. Yao, Simultaneous impedance measurements of two one-face sealed resonating piezoelectric quartz crystals for in situ monitoring of electrochemical processes and solution properties, *Analytica Chimica Acta,* Vol. 533, 2005, pp. 213-224.

[236]. H. He, Q. Xie, Y. Zhang, S. Yao, A simultaneous electrochemical impedance and quartz crystal microbalance study on antihuman immunoglobulin G adsorption and human immunoglobulin G reaction, *Journal of Biochemical and Biophysical Methods,* Vol. 62, 2005, pp. 191-205.

[237]. Q. Xie, Y. Zhang, M. Xu, Z. Li, Y. Yuan, S. Yao, Combined quartz crystal impedance and electrochemical impedance measurements during adsorption of bovine serum albumin onto bare and cysteine- or thiophenol- modified gold electrodes, *Journal of Electroanalytical Chemistry,* Vol. 478, 1999, pp. 1-8.

[238]. Q. Xie, C. Xiang, Y. Yuan, Y. Zhang, L. Nie, S. Yao, A novel dual-impedance-analysis EQCM system-investigation of bovine serum

albumin adsorption on gold and platinum electrode surfaces, *Journal of Colloid and Interface Science*, Vol. 262, 2003, pp. 107-115.

[239]. E. M. Pinto, D. M. Soares, C. M. Brett, Interaction of BSA protein with copper evaluated by electrochemical impedance spectroscopy and quartz crystal microbalance, *Electrochimica Acta*, Vol. 53, 2008, pp. 7460-7466.

[240]. F. Patolsky, M. Zayats, E. Katz, I. Willner, Precipitation of an insoluble product on enzyme monolayer electrodes for biosensor applications: characterization by faradaic impedance spectroscopy, cyclic voltammetry, and microgravimetric quartz crystal microbalance analyses, *Analytical Chemistry*, Vol. 71, 1999, pp. 3171-3180.

[241]. A. Sabot, S. Krause, Simultaneous quartz crystal microbalance impedance and electrochemical impedance measurements. Investigation into the degradation of thin polymer films, *Analytical Chemistry*, Vol. 74, 2002, pp. 3304-3311.

[242]. E. Briand, M. Zäch, S. Svedhem, B. Kasemo, S. Petronis, Combined QCM-D and EIS study of supported lipid bilayer formation and interaction with pore-forming peptides, *The Analyst*, Vol. 135, 2010, pp. 343-350.

[243]. A. S. Kliger, Maintaining safety in the dialysis facility, *Clinical journal of the American Society of Nephrology*, Vol. 10, 2015, pp. 688-695.

[244]. R. A. Ward, Avoiding toxicity from water-borne contaminants in hemodialysis: new challenges in an era of increased demand for water, *Advances in Chronic Kidney Disease*, Vol. 18, 2011, pp. 207-213.

[245]. R. A. Ward, Worldwide water standards for hemodialysis, *Hemodialysis International*, Vol. 11, 2007, pp. S18-S25.

[246]. A. D. Coulliete, M. J. Arduino, Hemodialysis and water quality, *Seminars in Dialysis*, Vol 26, 2013, pp. 427-438.

[247]. A. A. Fendley, R. A. Ward, Dialysate quality: New standards require a new approach to compliance, *Seminars in Dialysis*, Vol. 25, 2012, pp. 511-515.

[248]. T. Li, S. Dong, E. Wang, Label-free colorimetric detection of aqueous mercury ion (Hg2 +) using Hg2 +-modulated G-quadruplex-based DNAzymes, *Analytical Chemistry*, Vol. 81, 2009, pp. 2144-2149.

[249]. Z. Lin, X. Li, H.-B. Kraatz, Impedimetric immobilized DNA-based sensor for simultaneous detection of Pb2 +, Ag +, and Hg2 +, *Analytical Chemistry*, Vol. 83, 2011, pp. 6896-6901.

[250]. C.-W. Liu, C.-C. Huang, H.-T. Chang, Highly selective DNA-based sensor for lead(II) and mercury(II) ions, *Analytical Chemistry*, Vol. 81, 2009, pp. 2383-2387.

[251]. L. Chen, Z.-N. Chen, A multifunctional label-free electrochemical impedance biosensor for Hg2 +, adenosine triphosphate and thrombin, *Talanta*, Vol. 132, 2015, pp. 664-668.

[252]. C.-K. Chiang, C.-C. Huang, C.-W. Liu, H.-T. Chang, oligonucleotide-based fluorescence probe for sensitive and selective detection of

mercury(II) in aqueous solution, *Analytical Chemistry*, Vol. 80, 2008, pp. 3716-3721.

[253]. R. Zhang, W. Chen, Nitrogen-doped carbon quantum dots: Facile synthesis and application as a "turn-off" fluorescent probe for detection of Hg2 + ions, *Biosensors &Bioelectronics*, Vol. 55, 2014, pp. 83-90.

[254]. Z.-H. Gao, Z.-Z. Lin, X.-M. Chen, Z.-Z. Lai, Z.-Y. Huang, Carbon dots-based fluorescent probe for trace Hg2 + detection in water sample, *Sensors and Actuators B: Chemical*, Vol. 222, 2016, pp. 965-971.

[255]. F. Yan, D. Shi, T. Zheng, K. Yun, X. Zhou, L. Chen, Carbon dots as nanosensor for sensitive and selective detection of Hg2 + and l-cysteine by means of fluorescence "Off-On" switching, *Sensors and Actuators B: Chemical*, Vol. 224, 2016, pp. 926-935.

[256]. E. S. Childress, C. A. Roberts, D. Y. Sherwood, C. L. M. LeGuyader, E. J. Harbron, Ratiometric fluorescence detection of mercury ions in water by conjugated polymer nanoparticles, *Analytical Chemistry*, Vol. 84, pp. 1235-1239.

[257]. G.-H. Chen, W.-Y. Chen, Y.-C. Yen, C.-W. Wang, H.-T. Chang, C.-F. Chen, Detection of mercury(II) ions using colorimetric gold nanoparticles on paper-based analytical devices, *Analytical Chemistry*, Vol. 86, 2014, pp. 6843-6849.

[258]. W. Tang, D. B. Chase, D. L. Sparks, J. F. Rabolt, Selective and quantitative detection of trace amounts of mercury(II) ion (Hg2 +) and copper(II) ion (Cu2 +) using surface-enhanced Raman scattering (SERS), *Applied Spectroscopy*, Vol. 69, 2015, pp. 843-849.

[259]. L. Shi, G. Liang, X. Li, X. Liu, Impedimetric DNA sensor for detection of Hg2 + and Pb2 +, *Analytical Methods*, Vol. 4, 2012, pp. 1036-1040.

[260]. S. O. Kelley, J. K. Barton, DNA-mediated electron transfer from a modified base to ethidium: π-stacking as a modulator of reactivity, *Chemistry and Biology*, Vol. 5, 1998, pp. 413-425.

[261]. H. Torigoe, Y. Miyakawa, A. Ono, T. Kozasa, Thermodynamic properties of the specific binding between Ag + ions and C:C mismatched base pairs in duplex DNA, *Nucleosides Nucleotides Nucleic Acids*, Vol. 30, 2011, pp. 149-167.

[262]. C.-W. Liu, C.-C. Huang, H.-T. Chang, Highly selective DNA-based sensor for lead(II) and mercury(II) ions, *Analytical Chemistry*, Vol. 81, 2009, pp. 2383-2387.

[263]. H. Gong, X. Li, Y-type, C-rich DNA probe for electrochemical detection of silver ion and cysteine, *Analyst*, Vol. 136, 2011, pp. 2242-2246.

[264]. L. Chen, Z. Chen, A multifunctional label-free electrochemical impedance biosensor for Hg2 +, adenosine triphosphate and thrombin, *Talanta*, Vol. 132, 2015, pp. 664-668.

[265]. G. Mor-Piperberg, R. Tel-Vered, J. Elbaz, I. Willner, Nanoengineered electrically contacted enzymes on DNA scaffolds: Functional assemblies for the selective analysis of Hg2 + ions, *Journal of the American Chemical Society*, Vol. 132, 2010, pp. 6878-6879.

[266]. Z.-P. Yang, C.-J. Zhang, Designing of MIP-based QCM sensor for the determination of Cu(II) ions in solution, *Sensors and Actuators B: Chemical*, Vol. 142, 2009, pp. 210-215.

[267]. L. Santore, M. Barbaglio, L. Borgese, E. Bontempi, Polymer-grafted QCM chemical sensor and application to heavy metal ions real time detection, *Sensors & Actuators B*, Vol. 155, 2011, pp. 539-544.

[268]. C.-Y. Shen, Y.-M. Lin, R.-C. Hwang, Detection of Cu(II) ion in water using a quartz crystal microbalance, *Journal of Electrical and Electronic Engineering*, Vol. 4, 2016, pp. 13-17.

[269]. J. S. Metcalf, G. A. Codd, Analysis of cyanobacterial toxins by immunological methods, *Chemical Research in Toxicology*, Vol. 16, 2003, pp. 103-112.

[270]. M. G. Weller, Immunoassays and biosensors for the detection of cyanobacterial toxins in water, *Sensors*, Vol. 13, 2013, pp. 15085-15112.

[271]. F. Liu, A. N. Nordin, F. Li, I. Voiculescu, Lab-on-chip cell-based biosensor for label-free sensing of water toxicants, *Lab on a Chip*, Vol. 14, 2014, pp. 1270-1280.

[272]. Y. Xia, J. Zhang, L. Jiang, A novel dendritic surfactant for enhanced microcystin-LR detection by double amplification in a quartz crystal microbalance biosensor, *Colloids and Surfaces B: Biointerfaces*, Vol. 86, 2011, pp. 81-86.

[273]. M. V. Voinova, The theory of acoustic sensors application in air quality control, *Urban Climate*, Vol. 24, 2018, pp. 264-275.

Chapter 4

Solid State Colorimetric Biosensors

Dr. Nokuthula Ngomane

4.1. Introduction

Over the past decades, tremendous advances have been made in the development of sensors for a variety of analytes. A sensor is a device whose purpose is to sense and respond to characteristics of its environment such as light, heat, motion, moisture and pressure. Amongst other types of sensors, chemical and biological sensors (the latter is also known as a biosensor) continue to be significant subjects of research. Sensors find applications in very crucial areas including but not limited to clinical diagnosis and environmental health studies. A biosensor is an analytical device incorporating a biological sensing element including enzymes, nucleic acids, antibodies and macro-organisms for definite biological interaction [1, 2]. On the other hand, a chemical sensor can be defined as a device which responds to a specific analyte in a precise way through a chemical reaction and can be used for the qualitative or quantitative determination of the analyte [3]. A chemical sensor is concerned with sensing and measuring a specific chemical substance or a group of chemicals. For both chemical and biosensing, several detection (sensing) methods such as electrochemical, fluorescent and colorimetric methods can be employed [4-8]. Of these, colorimetric detection has attracted great interest because it is the easiest and the most convenient method because detection can be achieved with an unaided eye. Moreover, it is low cost and does not require the use of sophisticated equipment that need to be operated by technically competent personnel.

Nokuthula Ngomane
School of Chemical and Physical Sciences, University of Mpumalanga, South Africa

It can therefore be applied to field analysis and for point of care diagnosis.

Since colorimetric detection relies on colour changes of a component that is being investigated and compared with a colour of a standard by a bare eye, transforming the detection events into colour would be the fundamental challenge for colorimetric chemical and biological sensing. To date, colorimetric reagents such as gold nanoparticles (AuNPs), cerium oxide nanoparticles, carbon nanotubes, graphene oxide, conjugated polymers (e.g. polydiacetylene) and magnetic nanoparticles have been employed [9-15]. Although most of the colorimetric methods for biological and chemical analytes developed using these material gave a straightforward observation through simple readout schemes, often with high selectivity and sensitivity, most of them utilized solution-state sensing. Such colorimetric sensors in general still have a number of limitations which deter their widespread application. For instance, while they work well with clear samples, it would be a challenge to distinguish colour hue or intensity changes when analysing coloured samples or complex matrices (for example, food samples and blood samples). Hence solid-state colorimetric sensors which are portable, stable, and amenable to large scale fabrication are required. These are obtained by immobilizing the colorimetric reagents in solid materials such as gels, nanofibers (e.g. electrospun nanofibers), glass substrates, polymer films and polymer particles (e.g. molecularly imprinted polymers). The intention of this chapter is to reflect on the advances of solid-state colorimetric biosensors. The subsections are divided in terms of the solid support material used.

4.2. Polymer Films and Glass Substrates

The use of glass substrates and polymer films have enable easy fabrication of solid state colorimetric sensors. Poly(dimethylsiloxane) (PDMS) is one of the most commonly used polymeric material utilized in fabricating microfluidic devices. This is because the surface of PDMS can be easily modified chemically or physically masked to generate anionic, neutral and cationic surface. Taking advantage of the characteristics of this polymer material, Zhang and colleagues fabricated a PDMS-AuNPs composite film based biosensor [16]. Owing to the high stability of the composite film with shelf life of several months while the quantities of the reagents required to make it are in the μ range, Wu *et al.* have adopted it in conjunction with silver to produce a sensor for enhanced colorimetric detection of cardiac troponin I (Fig. 4.1) [17].

The intrinsic properties of smart materials such as AuNPs have been loaded into poly(oligo (ethylene glycol)methacrylate) brushes grown on glass for colorimetric sensing of Pb^{2+}. Langmuir-Blodgett films are one of the many useful tools in the design of solid state colorimetric sensors. Recently, a biosensor based on surface host-guest recognition that result in observable colour change upon addition of biotin-streptavidin on biotin functionalized Langmuir-Blodgett films of PDA was described [18].

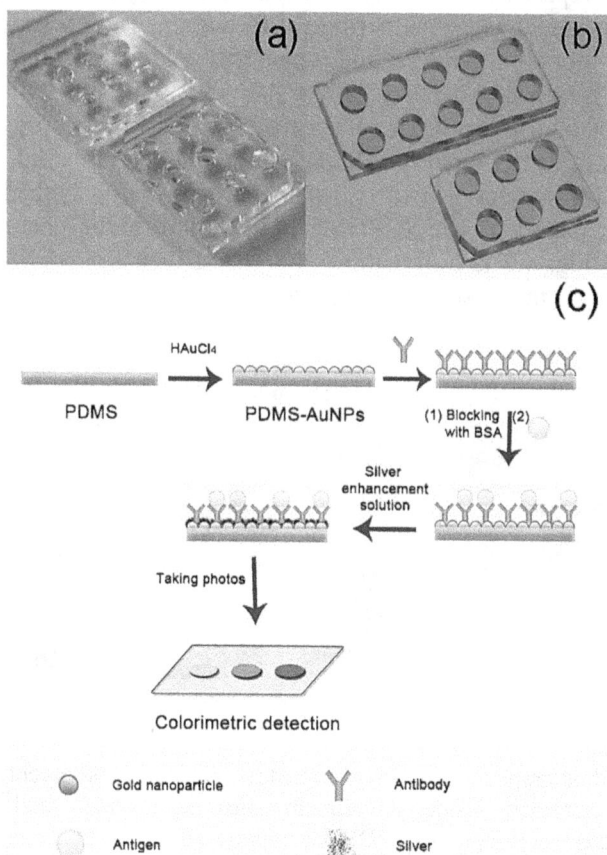

Fig. 4.1. The synthesis procedure for silver enhancement colorimetric detection of cardiac troponin I; (a) is the PDMS chip with $HAuCl_4$ solution; (b) is the photo of PDMS-AuNPs composite film; (c) schematic diagram for colorimetric detection [17].

135

4.3. Gels

Gels such as sol and hydrogels are applicable in the establishment of biosensors. This is due to their ability to preserve biological activity of the sensing materials encapsulated in them and they also allow for a wide variety of substrates to be embedded in high density. Hydrogels have high loading capacity. As a result, only very small amounts could be used and its patterning inside surface modified microfluidic channels could provide stable systems for fabrication of gel-based solid state biosensor. The use of hydrogel to produce portable sensors has been studied. For example, Baeissa et al. demonstrated a generable hydrogel based sensor for DNA prepare by covalently functionalizing acrydite-modified DNA to polyacrylamide hydrogel during gel formation. Their sensor was in a form of a monolith hydrogel and it changed colour from transparent to red in the presence of the target DNA [19]. In another illustration, Lee et al. reported on a colorimetric sensor for α-cyclodextrin employing PDA which was incorporated into poly(ethylene glycol) based hydrogel during photo-polymerization and the hydrogel matrix embedded with PDA was patterned within PDMS-based microfluidic channels [20]. Shown in Fig. 4.2 is a miniaturized form of their sensor whose main limitation was the slow response time.

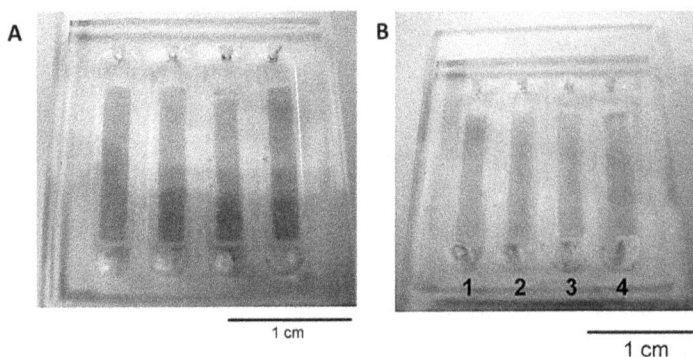

Fig. 4.2. Photographs of (A) PDA-embedded poly(ethylene glycol) dicrylate hydrogel patterned inside polydimethylsiloxane (PDMS) and reactions of hydrogel matrices in PDMS microchannels with concentrations of α-cyclodextrin solution after 120 min [20].

An earlier report by Gill and Ballesteros showed that a gel based sensor that is highly responsive (e.g. 1.1-1.6 min) can be obtained by carefully choosing reagents for the sensor and optimization. In their study, they fabricated a sensor by encapsulating PDA-phospholipid vesicles modified with immunoglobulin in hybrid sol-gel materials combined with silica and functionalized siloxanes. This chemical cocktail can be used to produce solid state sensor in the form of a monolith, thick films and microarrays [21]. Another example of a gel based sensor was described by Choodum and co-workers [22]. In their studies, they have utilised the sensor to quantify the drug methamphetamine in illicit tablets and urine sample.

4.4. Molecularly Imprinted Polymers

Molecularly imprinted polymers (MIPs) are designed such that they mimic the natural receptors. The molecular imprinting process produces polymers with specific recognition sites for the targeted analyte. As such, they are suitable candidates for application in the fabrication of sensors. MIPs have been combined with various chromophores to make colorimetric sensors that served their purpose in a satisfactory manner. Reports for such include the sensor for phenol employing MIP membranes that formed a coloured complex with 4-aminoantipyrine in alkaline solution in the presence of ferricyanide [23]. The MIP sensor showed high selectivity towards phenol and relatively low binding of its structural analogues. In addition, it was characterized by a wide detection range (50 nM-10 mM) with a low detection limit of 50 nM. Other MIP based colorimetric sensor have been developed for detection of atrazine [24], vanillin [25] and p-nitrophenol [26] among other compounds.

While literature is rich with reports about merging the special characteristics of other various chromophores, there are very few reports on the study of colorimetric sensors based on MIPs that integrate the molecular imprinting technique with metal NPs. To date, there are less than five such studies that have been reported. Amongst the few, Matsui and co-workers reported a MIP with immobilized 11-mercaptoundecanoic acid stabilized AuNPs as colorimetric sensor material for detecting adrenaline [27]. The MIP was utilized as both a solid support and for selectivity. The method is based on the swelling of the MIP during analyte rebinding, which result in colour change. The reaction scheme is shown in Fig. 4.3.

The latest study reported in this area was by Sergeyeva and co-workers [23]. The authors proposed a sensor for selective detection of phenol

137

based on free-standing MIP membranes. When the phenol is adsorbed by the MIP membranes and reacted with 4-aminoantipyrine, a pink product is obtained. The intensity of the pink colour was found to be proportional to the amount of phenol present in the analysed sample.

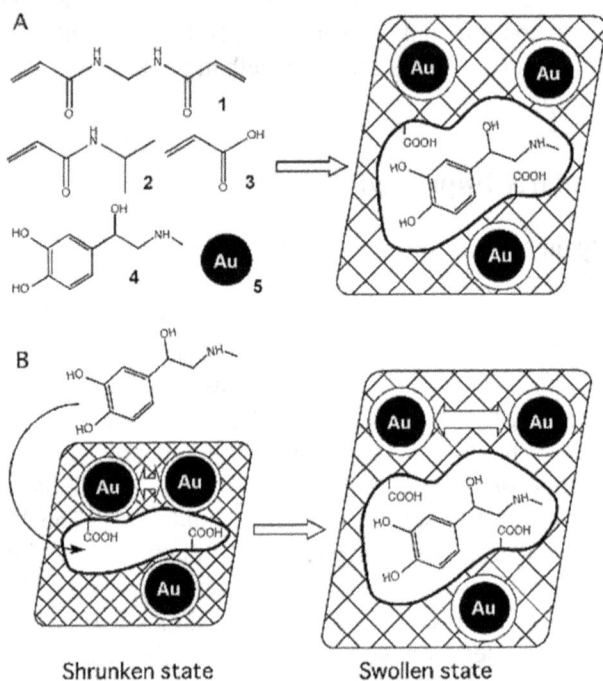

Shrunken state   Swollen state

Fig. 4.3. Schematic representation of the preparation of a MIP with immobilized AuNPs (A) and the detection of an analyte upon selective swelling of the imprinted polymer (B) [27].

4.5. Nanofibers

Recently, the electrospinning technique has attracted great interest in the synthesis of polymer nanofibers and two-dimensional spider-web-like nano-nets. Electrospinning is a low cost technique for making polymer nanofibers that have ultrahigh surface area to volume ratio and controllable structure and diameters. The properties of the synthesized nanofibers are achieved by manipulating the electrospinning conditions such as the viscosity of the polymer, accelerating voltage, distance from the tip of the needle to collector, and the flow rate [28, 29]. The recently discovered architecture, the nano-nets, can create enhanced interconnectivity and extra specific surface area and enable the diffusion of analytes into the nanofiber/nano-nets membranes. Therefore, they are attractive candidates as sensing materials loaded with colorimetric reagents and receptors for ultrasensitive sensors [30-32].

The mechanical and physical strength of nylon-6 fibers combined with the intrinsic characteristics of metal nanoparticles normally give materials with exceptional properties. Colorimetric sensors composed of these materials for sensing of biological and environmental compounds have been developed. For example, utilizing electrospun polyamide-6 (nylon-6)/nitrocellulose nanofiber/nano-nets membranes embedded with bovine serum decorated AuNPs (a colorimetric reagent), Li *et al.* fabricated a selective, facile and sensitive (with low detection limit of 0.2 µM) sensor strip for Pb^{2+} [30]. The stability of the bovine decorated AuNPs was enhanced by the extremely large and specific surface areas and high porosity of the membranes. The colorimetric detection event involved a visual colour change of the strips from deep pink to white after introduction to Pb^{2+} liquor. Together with my colleagues, we have developed a colorimetric probe for dopamine detection using un-functionalized AuNP hosted in nylon-6 fibers [9]. The purple nanofiber turns navy blue/black in the presence of dopamine.

Metal alloys have also been utilized as colorimetric probes. Ondigo *et al.* reported a method for colorimetric sensing of nickel(II) using glutathione stabilized silver-copper (Ag-Cu) alloy naoparticles embedded in electropun nanofibers [33]. The sensor selectively detected Ni^{2+} in the presence of selected transition metals and its sensitivity effect towards other alkali and alkali earth metal ions was found to be negligible. In another study, a colorimetric sensor based on copper-gold alloy nanoparticles employing AuNPs supported in electrospun nylon-6 nanofiber for detection of ascorbic acid (AA) was described [34]. The white nanofiber prepared by electrospinning undergo a colour change from white to blue due to the aggregation of the alloy nanoparticles showing selectivity for AA in the presence of interfering species at pH 2-7. Also using nylon-6 nanofibres, Matteo *et al.* demonstrated a sensor for reducing sugars in which gold salt was impregnated within the polymer matrix instead of AuNPs [35]. On interaction with the reducing sugars in alkaline media, the gold salts formed AuNPs thereby resulting in a colour change from white to purple. Other polymer nanofiber have also been used as solid support for nanoparticles in the development of electrospun based solid state colorimetric sensors [36].

Conjugated polymers such as polydiacetylene (PDA) and polyaniline have been incorporated with nanofibers to make solid state colorimetric sensors. PDA is well known for its distinct colour changes from red to blue due to external stimuli such as pH, temperature and other mechanical stress caused by changes in the conjugated (ene-yne) PDA backbone [37, 38]. Chemical sensing of volatile organic compounds [39], hydrochloric gas [40] have been realized utilizing PDAs hosted in nanofibers. The latter was prepared by mixing PDA monomers bearing trimethyl amine with poly(ethylene oxide) into electrospun nanofibers. The nanofiber mat change colour from blue to red in the presence of hydrochloric acid gas instead of the widely known red to blue (Fig. 4.4).

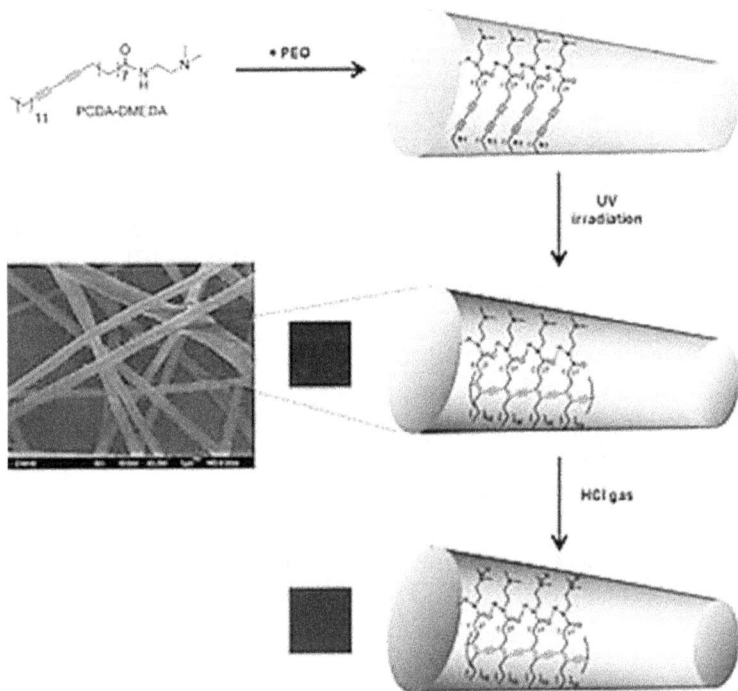

Fig. 4.4. Schematic representation of the preparation of PDA-embedded electrospun nanofiber [40].

Polyaniline (PANI) is recognized to have colorimetric transitions when it is exposed to certain metals and has been impregnated in solid substrates for the purpose of making sensors [41, 42]. An example of some of the solid materials was reported by Ding *et al.* [43] where they prepared a blend of nylon-6 and PANI via electrospinning. The nanofiber or net obtained was evaluated for the detection of Cu^{2+} in water and a visual colour change from white to blue was observed.

4.6. Other Substrates

Incorporating colorimetric reagents with materials such as cellulose or paper substrate can result in devices that are portable, easy to use less expensive but still performs exceptional in terms of the colorimetric response time and specificity. Cellulose, a naturally occurring polymer has been used as a solid support for 10,12-pentacosadyinoic acid+N-[(2-tetradecanamide)-ethyl]-ribonamide vesicle to colorimetrically determine a foodborne bacteria. Dipstick based colorimetric materials using AuNPs have recently caught the interest of a number of researchers as they allow development of novel sensors based on dissociation of agglomerated AuNPs. Liu *et al.* have employed aggregates of AuNPs immobilized on a lateral flow device to illustrate a dipstick assay in which the nanoparticle aggregates are broken into red coloured dispersed AuNPs upon addition of target analytes in flow buffer [44]. Also using the lateral flow device, Debapriya and co-workers [45] developed a dipstick test for detection of Pb^{2+} in paint utilizing AuNPs DNAzymes conjugates. In another study a sensor that also relied on the dispersion of aggregates of AuNPs was demonstrated on paper [46]. Two types of paper substrate namely hydrophobic and hydrophilic paper were used and they were both found suitable for the biosensing of endonuclease I and adenosine. A schematic representation of the sensing process is depicted in Fig. 4.5. Fixing of dyes on a filter paper for solid state sensing has also been explored [47].

4.7. Conclusions

There is a growing demand for biosensors that are cost effective, sensitive and easy to use. Indeed, the use of solid state colorimetric sensors has proved to be beneficial; it is now a growing area of research. The merits of such sensors are appreciated more when the colours are further analyzed using image processing software such as ImageJ; more especially in cases where the colour changes are not so clear or for visually impaired people. With the advancing technology, this program could be added as an App into cellphones and any one owning a smart phone will be able to quantify the results.

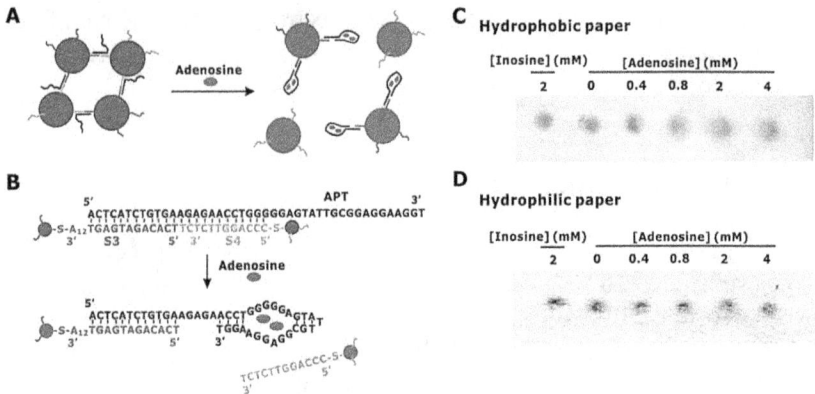

Fig. 4.5. (A) Schematic illustration of an adenosine colorimetric sensor using AuNP aggregates. Adenosine-binding DNA aptamer (APT) is used as a cross-linker to bridge AuNPs attached with cDNA strands (S3 and S4). (B) The sequences of the DNA molecules used. The addition of adenosine which binds to and dissociates the aptamer molecule from the cDNA strand leads to the redispersion of AuNP aggregates. (C) and (D) are adenosine-sensing assays on hydrophobic and hydrophilic paper, respectively. One µL of analyte solution was applied, and all images were obtained after 1 min [46].

References

[1]. C. Bosch-Orea, M. Farré, D. Barceló, Biosensors and bioassays for environmental monitoring, *Comprehensive Analytical Chemistry*, Vol. 77, 2017, pp. 337-338.

[2]. Shruthi G.S., Amitha C.V., Mathew B.B., Biosensors: A modern day achievement, *Journal of Instrumentation Technology*, Vol. 2, 2014, pp. 26-39.

[3]. S. Shanmugam, Enzyme Technology, *IK International Pvt Ltd*, 2009.

[4]. D. Nidzworski, P. Pranszke, M. Grudniewska, E. Król, B. Gromadzka, Universal biosensor for detection of influenza virus, *Biosensors and Bioelectronics*, Vol. 59, 2014, pp. 239-242.

[5]. D. S. Campos-Ferreira, E. V. M. Souza, G. A. Nascimento, D. M. L. Zanforlin, M. S. Arruda, M. F. S. Beltrão, A. L. Melo, D. Bruneska, J. L. Lima-Filho, Electrochemical DNA biosensor for the detection of human papillomavirus E6 gene inserted in recombinant plasmid, *Arabian Journal of Chemistry*, Vol. 9, 2016, pp. 443-450.

[6]. S. M. Taghdisi, N. M. Danesh, P. Lavaee, A. Sarreshtehdar Emrani, M. Ramezani, K. Abnous, Aptamer biosensor for selective and rapid determination of insulin, *Analytical Letters*, Vol. 48, 2015, pp. 672-681.

[7]. Y. Zhou, J. F. Zhang, J. Yoon, Fluorescence and colorimetric chemosensors for fluoride-ion detection, *Chemical Reviews*, Vol. 114, 2014, pp. 5511-5571.

[8]. C. F. Wan, J. L. Chir, A. T. Wu, A colorimetric sensor for the selective detection of fluoride ions, *Luminescence*, Vol. 32, 2017, pp. 353-357.

[9]. N. Ngomane, N. Torto, R. Krause, S. Vilakazi, A colorimetric probe for dopamine based on gold nanoparticles-electrospun nanofibre composite, *Materials Today: Proceedings*, Vol. 2, 2015, pp. 4060-4069.

[10]. M. I. Kim, K. S. Park, H. G. Park, Ultrafast colorimetric detection of nucleic acids based on the inhibition of the oxidase activity of cerium oxide nanoparticles, *Chemical Communications*, Vol. 50, 2014, pp. 9577-9580.

[11]. S. X. Zhang, S. F. Xue, J. Deng, M. Zhang, G. Shi, T. Zhou, Polyacrylic acid-coated cerium oxide nanoparticles: An oxidase mimic applied for colorimetric assay to organophosphorus pesticides, *Biosensors and Bioelectronics*, Vol. 85, 2016, pp. 457-463.

[12]. Y. Huang, J. Liu, M. Zou, Q. Zhang, J. Chao, W. Zhao, D. Wu, S. Su, L. Wang, Colorimetric detection and efficient monitoring of a potential biomarker of lumbar disc herniation using carbon nanotube-based probe, *Science China Chemistry*, Vol. 59, 2016, pp. 493-496.

[13]. E. J. Son, J. S. Lee, M. Lee, C. H. T. Vu, H. Lee, K. Won, C. B. Park, Self-adhesive graphene oxide-wrapped TiO2 nanoparticles for UV-activated colorimetric oxygen detection, *Sensors and Actuators B: Chemical*, Vol. 213, 2015, pp. 322-328.

[14]. Q. Xu, S. Lee, Y. Cho, M. H. Kim, J. Bouffard, J. Yoon, Polydiacetylene-Based Colorimetric and Fluorescent Chemosensor for the Detection of Carbon Dioxide, *Journal of the American Chemical Society*, Vol. 135, 2013, pp. 17751-17754.

[15]. N. Duan, B. Xu, S. Wu, Z. Wang, Magnetic nanoparticles-based aptasensor using gold nanoparticles as colorimetric probes for the detection of salmonella typhimurium, *Analytical Sciences*, Vol. 32, 2016, pp. 431-436.

[16]. Q. Zhang, J.-J. Xu, Y. Liu, H.-Y. Chen, In-situ synthesis of poly(dimethylsiloxane)-gold nanoparticles composite films and its application in microfluidic systems, *Lab on a Chip*, Vol. 8, 2008, pp. 352-357.

[17]. W. Y. Wu, Z. P. Bian, W. Wang, J. J. Zhu, PDMS gold nanoparticle composite film-based silver enhanced colorimetric detection of cardiac troponin I, *Sensors and Actuators B: Chemical*, Vol. 147, 2010, pp. 298-303.

[18]. E. Geiger, P. Hug, B. A. Keller, Chromatic transitions in polydiacetylene Langmuir-Blodgett films due to molecular recognition at the film surface studied by spectroscopic methods and surface analysis, *Macromolecular Chemistry and Physics*, Vol. 203, 2002, pp. 2422-2431.

[19]. A. Baeissa, N. Dave, B. D. Smith, J. Liu, DNA-functionalized monolithic hydrogels and gold nanoparticles for colorimetric DNA detection, *Applied Materials & Interfaces*, Vol. 2, 2010, pp. 3594-3600.

[20]. N. Y. Lee, Y. K. Jung, H. G. Park, On-chip colorimetric biosensor based on polydiacetylene (PDA) embedded in photopolymerized poly(ethylene glycol) diacrylate (PEG-DA) hydrogel, *Biochemical Engineering Journal*, Vol. 29, 2006, pp. 103-108.

[21]. I. Gill, A. Ballesteros, Immunoglobulin-polydiacetylene sol-gel nanocomposites as solid-state chromatic biosensors, *Angewandte Chemie International Edition in English*, Vol. 42, 2003, pp. 3264-3267.

[22]. A. Choodum, P. Kanatharana, W. Wongniramaikul, N. NicDaeid, A sol-gel colorimetric sensor for methamphetamine detection, *Sensors and Actuators B: Chemical*, Vol. 215, 2015, pp. 553-560.

[23]. T. A. Sergeyeva, D. S. Chelyadina, L. A. Gorbach, O. O. Brovko, E. V. Piletska, S. A. Piletsky, L. M. Sergeeva, A. V. El'skaya, Colorimetric biomimetic sensor systems based on molecularly imprinted polymer membranes for highly-selective detection of phenol in environmental samples, *Biopolymers and Cell*, Vol. 30, 2014, pp. 209-215.

[24]. Z. Wu, C. A. Tao, C. Lin, D. Shen, G. Li, Label-free colorimetric detection of trace atrazine in aqueous solution by using molecularly imprinted photonic polymers, *Chemistry – A European Journal*, Vol. 14, 2008, pp. 11358-11368.

[25]. H. Peng, S. Wang, Z. Zhang, H. Xiong, J. Li, L. Chen, Y. Li, Molecularly imprinted photonic hydrogels as colorimetric sensors for rapid and label-free detection of vanillin, *Journal of Agricultural and Food Chemistry*, Vol. 60, 2012, pp. 1921-1928.

[26]. F. Xue, Z. Meng, Y. Wang, S. Huang, Q. Wang, W. Lu, M. Xue, A molecularly imprinted colloidal array as a colorimetric sensor for label-free detection of p-nitrophenol, *Analytical Methods*, Vol. 6, 2014, pp. 831-837.

[27]. J. Matsui, K. Akamatsu, S. Nishiguchi, D. Miyoshi, H. Nawafune, K. Tamaki, N. Sugimoto, Composite of Au nanoparticles and molecularly imprinted polymer as a sensing material, *Analytical Chemistry*, Vol. 76, 2004, pp. 1310-1315.

[28]. Q. Shi, N. Vitchuli, L. Ji, J. Nowak, M. McCord, M. Bourham, X. Zhang, A facile approach to fabricate porous nylon 6 nanofibers using silica nanotemplate, *Journal of Applied Polymer Science*, Vol. 120, 2011, pp. 425-433.

[29]. Q. Shi, N. Vitchuli, J. Nowak, J. Noar, J. M. Caldwell, F. Breidt, M. Bourham, M. McCord, X. Zhang, One-step synthesis of silver nanoparticle-filled nylon 6 nanofibers and their antibacterial properties, *Journal of Material Chemistry*, Vol. 21, 2011, pp. 10330-10335.

[30]. Y. Li, Y. Si, X. Wang, B. Ding, G. Sun, G. Zheng, W. Luo, J. Yu, Colorimetric sensor strips for lead (II) assay utilizing nanogold probes immobilized polyamide-6/nitrocellulose nano-fibers/nets, *Biosensors and Bioelectronics*, Vol. 48, 2013, pp. 244-250.

[31]. B. Ding, M. Wang, X. Wang, J. Yu, G. Sun, Electrospun nanomaterials for ultrasensitive sensors, *Materials Today*, Vol. 13, 2010, pp. 16-27.

[32]. X. Wang, B. Ding, J. Yu, M. Wang, Highly sensitive humidity sensors based on electro-spinning/netting a polyamide 6 nano-fiber/net modified by polyethyleneimine, *Journal of Material Chemistry*, Vol. 21, 2011, pp. 16231-16238.

[33]. D. Ondigo, B. Mudabuka, B. Pule, Z. Tshentu, N. Torto, A colorimetric probe for the detection of Ni2+ in water based on Ag-Cu alloy nanoparticles hosted in electrospun nanofibres, *Water SA*, Vol. 42, 2016, pp. 408-414.

[34]. B. Mudabuka, D. Ondigo, S. Degni, S. Vilakazi, N. Torto, A colorimetric probe for ascorbic acid based on copper-gold nanoparticles in electrospun nylon, *Microchimica Acta*, Vol. 181, 2014, pp. 395-401.

[35]. M. Scampicchio, A. Arecchi, S. Mannino, Optical nanoprobes based on gold nanoparticles for sugar sensing, *Nanotechnology*, Vol. 20, 2009, 135501.

[36]. B. O. Pule, S. Degni, N. Torto, Electrospun fibre colorimetric probe based on gold nanoparticles for on-site detection of 17^2-estradiol associated with dairy farming effluents, *Water SA*, Vol. 41, 2015, pp. 27-32.

[37]. Y. Song, W. Wei, X. Qu, Colorimetric biosensing using smart materials, *Advanced Materials*, Vol. 23, 2011, pp. 4215-4236.

[38]. Y. K. Jung, H. G. Park, J.-M. Kim, Polydiacetylene (PDA)-based colorimetric detection of biotin-streptavidin interactions, *Biosensors and Bioelectronics*, Vol. 21, 2006, pp. 1536-1544.

[39]. J. Yoon, S. K. Chae, J.-M. Kim, Colorimetric Sensors for Volatile Organic Compounds (VOCs) Based on Conjugated Polymer-Embedded Electrospun Fibers, *Journal of the American Chemical Society*, Vol. 129, 2007, pp. 3038-3039.

[40]. H. Jeon, J. Lee, M. H. Kim, J. Yoon, Polydiacetylene-based electrospun fibers for detection of HCl gas, *Macromolecular Rapid Communications*, Vol. 33, 2012, pp. 972-976.

[41]. G. Li, J. Zheng, X. Ma, Y. Sun, J. Fu, G. Wu, Development of QCM trimethylamine sensor based on water soluble polyaniline, *Sensors*, Vol. 7, 2007, pp. 2378-2388.

[42]. M. M. Ayad, G. El-Hefnawey, N. L. Torad, A sensor of alcohol vapours based on thin polyaniline base film and quartz crystal microbalance, *Journal of Hazardous Material*, Vol. 168, 2009, pp. 85-88.

[43]. B. Ding, Y. Si, X. Wang, J. Yu, L. Feng, G. Sun, Label-free ultrasensitive colorimetric detection of copper(II) ions utilizing polyaniline/polyamide-6 nano-fiber/net sensor strips, *Journal of Materials Chemistry*, Vol. 21, 2011, pp. 13345-13353.

[44]. J. Liu, D. Mazumdar, Y. Lu, A Simple and sensitive "dipstick" test in serum based on lateral flow separation of aptamer-linked nanostructures, *Angewandte Chemie International Edition*, Vol. 45, 2006, pp. 7955-7959.

[45]. D. Mazumdar, J. Liu, G. Lu, J. Zhou, Y. Lu, Easy-to-use dipstick tests for detection of lead in paints using non-cross-linked gold nanoparticle-DNAzyme conjugates(), *Chemical Communications*, Vol. 46, 2010, pp. 1416-1418.

[46]. W. Zhao, M. M. Ali, S. D. Aguirre, M. A. Brook, Y. Li, Paper-based bioassays using gold nanoparticle colorimetric probes, *Analytical Chemistry*, Vol. 80, 2008, pp. 8431-8437.

[47]. J. Lee, Z. U. Shin, G. Mavlonov, I. Abdurakhmonov, T.-H. Yi, Solid-Phase colorimetric method for the quantification of fucoidan, *Applied Biochemistry and Biotechnology*, Vol. 168, 2012, pp. 1019-1024.

Chapter 5

Sensing of H_2O_2 as Cancer Biomarker with Ldhs Nanostructures

Muhammad Asif, Ayesha Aziz, Zhengyun Wang and Hongfang Liu

5.1. Hydrogen Peroxide as Cancer Biomarker

In several physiological events, significant scientific evidences show that hydrogen peroxide (H_2O_2) is considered as a gesturing molecule for precise as well as prompt determination of oxidative stress that can be associated with different kinds of chronic diseases like Alzheimer's, atherosclerosis, lungs injury, cardiovascular diseases, parasitic infections, diabetes and cancer [1, 2]. It is not simply a by-product of numerous oxidases in various biological functions but also a requisite mediator in biomedical, pharmaceutical, food and environmental analysis [3, 4]. In living systems, its massive accumulation is detrimental for normal growth of cells and is engendered by the oxidative mitochondrial functions, incomplete reduction of oxygen and metabolic reactions occurring in live cells [5]. The reactive oxygen species play critical role in proliferation, physiological intracellular signaling transduction, steady development, abiotic anxiety influences, response to lethal attacks, relocation and distinction of healthy cell [6]. Nevertheless, appropriate level of H_2O_2 is incredibly significant as a progressive tracer of biochemical reactions but its excessive production in cellular environments that causes its penetration to other cellular compartments is extremely pathogenic to living organisms [7]. Recently, biomarker based cancer diagnosis has drawn extensive attention of scientists because of their high reliability, sensitivity and simplicity. It is worth mentioning that H_2O_2 plays a vital role in cell proliferation, cell

Hongfang Liu
Key Laboratory of Material Chemistry and Service Failure, Wuhan, China

death, response to chronic diseases and intracellular signaling transduction [1, 6]. A key feature of studying H_2O_2 in living organisms is to ensure that its concentration in cellular compartments should be in physiological range of 1 nM to 0.5 μM in all mammals [8, 9]. Studies have profiled the cellular functions associated with endogenous H_2O_2 efflux in living cells are emerging recognized biomarker for cancer diagnostics. Therefore, it is necessary to monitor the exact level of H_2O_2 that is an important pathway for reliable understanding of pathological, physiological and biomedical role of H_2O_2 [10]. Consequently, countless persistent attempts has been done to construct analytical techniques for quantifying H_2O_2 such as calorimetric, titrimetric [11] and chemiluminescence [12] and electrochemical sensors [13]. Among these strategies, electrochemical sensors have conquered pronounced significance because of their ultrasensitivity, selectivity, low cost and low detection limits.

5.2. LDHs Based Electrochemical Biosensors

The electrochemical biosensors have been advertised as a particularly promising class of imminent generation of analytical devices with intriguing characteristics because of incredible performance superior to conventional analytical tools since the successful demonstration of biosensor by Leland C. Clark [14]. Considerable attention has been devoted by numerous researchers to achieve improved functionality of this device for the detection of several small molecular metabolites. During past decade, electrochemical sensors and biosensors have conquered a wealth of interest because of facilitating multiplexed, simultaneous detection of various analytes at low concentration as well as exciting features like economical, simplicity, biocompatibility and excellent sensitivity [15]. With the ongoing gigantic demands of fast point-of-care medical strategies, biosensors are being extensively used in hereditary investigation, human immunodeficiency virus, angiogenesis, and metastasis [16, 17]. The *in vitro* sensing of biological molecules, cellular targets from blood serum, urine, honey, living systems and tumorigenic environments are considered beneficial over *in vivo* analysis [18] because of simplicity in environmental manipulation, minimum operational instrumentation, disposable, portable benefits [19], which are critically important for prompt on-site diagnosis of diseases. More recently, electrochemical sensors are considered as extremely sensitive platform for detecting various chemical and biochemical targets, which are suitably used in early cancer diagnostics.

However, extensively used enzyme based biosensors have been confined, because of several serious limitations like expensive, environmental instability [20], complicated immobilization and denaturation [21]. These obstacles can be controlled by exploring noble metal nanoparticles as substitutional catalysts for H_2O_2 sensing [22]. High cost, easily poisoned, toxic to environment and rare resources of noble metals reveal the perseverance to develop economical and excellent performance catalyst as non-enzymatic H_2O_2 detection practically. Graphene based CNT/MnO_2 nanocomposite papers [2, 23], growth of metal–metal oxide nanostructures on freestanding graphene paper [24] and graphene wrapped Cu_2O nanocubes [25] are the reported articles on graphene doped nanomaterials. Moreover, graphene having enormous intrinsic limitations like clumping in solution, costly, hydrophobic in nature, complicated and extended fabrication time, is certainly a big hindrance in assembling graphene doped nanomaterials for amperometric biosensors [26, 27].

Inorganic nanomaterials, a good competitor of noble metals, for instance Cu, Mn, Fe, and their oxides/hydroxides have fascinated the researchers to be used as biosensors [28] owing to their large specific surface area, fast electron transfer and profound electrocatalytic activities. Layered doubled hydroxides (LDHs) 2D material, having tunable configurations, also known as hydrotalcite like compounds (HTlcs), is considered a promising class of inorganic layered nanomaterials [29], with alternately distributed divalent and trivalent cations in sheets and charge balancing anions between brucite type host layers. By the virtue of these integrated structural features, LDHs have inevitably captured immense interest of scientists to utilize them in numerous fields like photoactive, electroactive, cancer therapy and electrochemical biosensor materials [30]. Furthermore, versatile distinctive characteristics coupled with LDH based nanomaterials depend upon multilamellar structure, memory effect, anion exchange feature, thermally stable, facile fabrication, high surface area, diversed oxides homogeneity and substantial electroactivity. Under the auspices of these distinct properties, LDHs are liable to use as a substitute conductive substrate of graphene in modifying electrode surface [31, 32]. Additionally, hierarchical LDH architectures with huge surface -OH groups, intrinsic positive charge, topotactic transformation [33], possess amazing intercalation features for entrapping metal nanoparticles [34]. Particularly, inorganic metal nanoparticles incorporated with LDHs have been captured much devotion of researchers to exploit them in electroanalytical chemistry and biosensor devices as shown in Fig. 5.1 [35]. These applications are valuable in scaling-up the potential of LDHs for physiological and pathological studies.

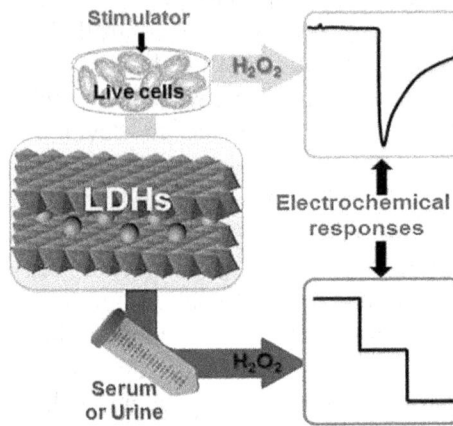

Fig. 5.1. Schematic illustration of LDHs materials functioning as biosensor devices.

5.3. Fabrication of Nanohybrids

To achieve the nanosized LDHs and heteroassembled LDH composite materials, various strategies can be of under consideration including adsorption, anion exchange within interlayer domain, delaminated restacking, restoration and directly co-precipitation. It is noted that soft synthetic pathways under few standard conditions like temperature, pressure, and pH are required to preserve LDHs composite structures. It is worth mentioning that LDHs have two interesting features which impart specific catalytic characteristics to LDH based materials. Among them: (i) the brucite-type layers capable of developing heterogeneous solid base catalyst owing to exchangeable anionic species within the interlayer domains; (ii) without segregated cations, the uniform distribution of metal cations at atomic-scale in the active transition metal cation that could stimulate marvelous catalytic brucite-type layers of LDHs possess catalytically abilities.

5.3.1. Simple Coprecipitation Method

CuO-MnAl nanostructures were prepared by facile co-precipitation and hydrothermal routs as reported earlier by our group (Fig. 5.2) [36]. Six different samples having fixed amount of MnAl (M^{2+}/M^{3+} atomic ratio of 3) with different amount of $Cu(CH_3COO)_2 \cdot H_2O$, were synthesized. Typically, an aqueous solutions A (50 mL) 0.1 M NaOH consisting varied amount of $Cu(CH_3COO)_2 \cdot H_2O$ (0.03-5 mM) stirred magnetically. After immediate appearance of blue $Cu(OH)_2$ precipitates, 1 mM ascorbic acid was added and stirred vigorously for 5 min. During this blue color of solution turned to orange, indicating the formation of CuO. Solution B (50 mL) containing 15 mM $MnCl_2.4H_2O$, 5 mM $Al(NO_3)_3.9H_2O$ was added drop wise into solution A at room temperature with constant stirring. The pH of the solution was sustained at 10 ± 0.02 by adding an alkaline solution of 2 M NaOH and 30 mM Na_2CO_3 during reaction. Then the product was transferred into teflon-lined stainless autoclave, sealed it and kept at $100\,^{\circ}C$ for 6 h. Then allowed to cool down naturally at room temperature and centrifuged. Then neutral pH of precipitates was obtained by washing several times with deionized water and final product was dried at 60 °C for overnight. As a contrast MnAl with Mn:Al atomic ratio 3 was also prepared under the same procedure. Finally, 5 µL suspension of 2 mg/mL of as synthesized CuO@MnAl was casted on the surface of polished glass carbon electrode (GCE) with alumina slurry and dried under ambient temperature.

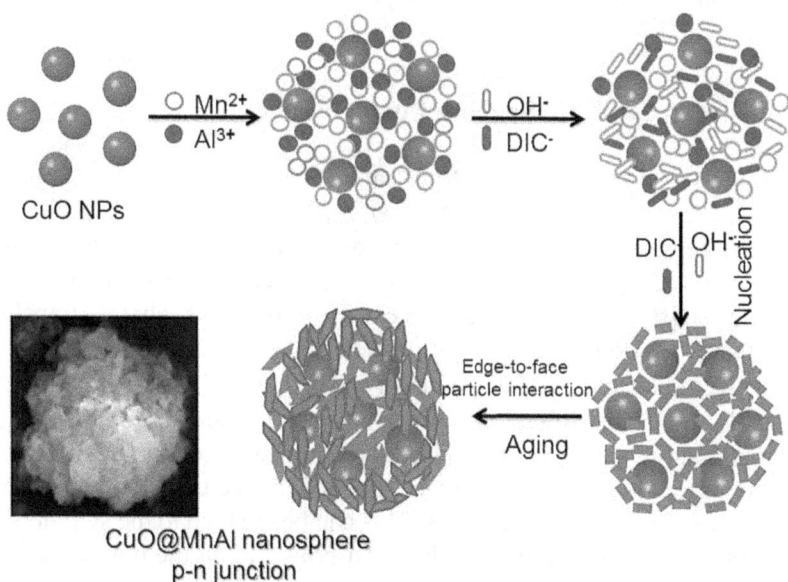

Fig. 5.2. Flow chart of the fabrication mechanism of CuO@MnAl NSs.

5.4. Cell Culture

Six kinds of living cells, i.e., epithelial normal cells HBL-100, breast cancer cells MCF-7, glioma brain cancer cells U87, HeLa cells, mouse hepatoma cell line H-22 and human hepatocellular carcinoma cell line BEL-7402 were purchased from Center for Tissue Engineering and Regenerative Medicine, Union Hospital, HUST (Wuhan, China). Cells were kept in Dulbecco's Modified Eagle Medium (DMEM) comprising of 10 % Fetal Bovine Serum, 100 units/mL penicillin and 100 μg/mL streptomycin in an incubator at 37 °C with 5 % CO_2 and 95 % humidified atmosphere and subcultured after every third day.

5.5. Amperometric Detection of Extracellular H_2O_2

The cells with 80 % confluency were seeded after centrifugation for amperometric H_2O_2 detection. The working electrode CuO-MnAl/GCE was placed near live cells in Dulbecco's Modified Eagle Medium (DMEM) at 37 °C for sensing H_2O_2 secretions. Amperometric feedback was recorded for H_2O_2 flux at CuO-MnAl modified GCE at -0.85 V upon the addition of 10 μM (0.1 mM) fMLP to cell suspension [37]. The cells are encouraged to evolve H_2O_2 by injecting fMLP.

5.6. Electrocatalytic Sensing Performance

The interfacial properties of surface modified by Cu3-MnAl and MnAl LDHs electrodes were evaluated with $[Fe(CN)_6]^{3-/4}$ redox probe to explore the intrinsic electrocatalytic properties. Fig. 5.3B shows the characteristic Nyquist plots, fitted with Randle equivalence circuit (inset of Fig. 5.3B) where R_s resistance of solution, R_{ct} charge transfer resistance, C_{dl} double layer capacitance and W is Warburg constant, of both electrodes in 0.1 M KCl solution consisting of 1 mM $[Fe(CN)_6]^{3-/4}$. Notably, the modification of MnAl substrate with integrated nanostructures of copper oxides leads to significant reduction in R_{ct} up to 480 Ω, which is indicative of huge mass and charge transfer capacity of Cu3-MnAl LDHs electrocatalysts. This can be credited to the symmetric distribution and substantial loading of n-type CuO in the host layers of p-type MnAl LDHs that provides compulsory conductive pathways for the passage of electrons [38].

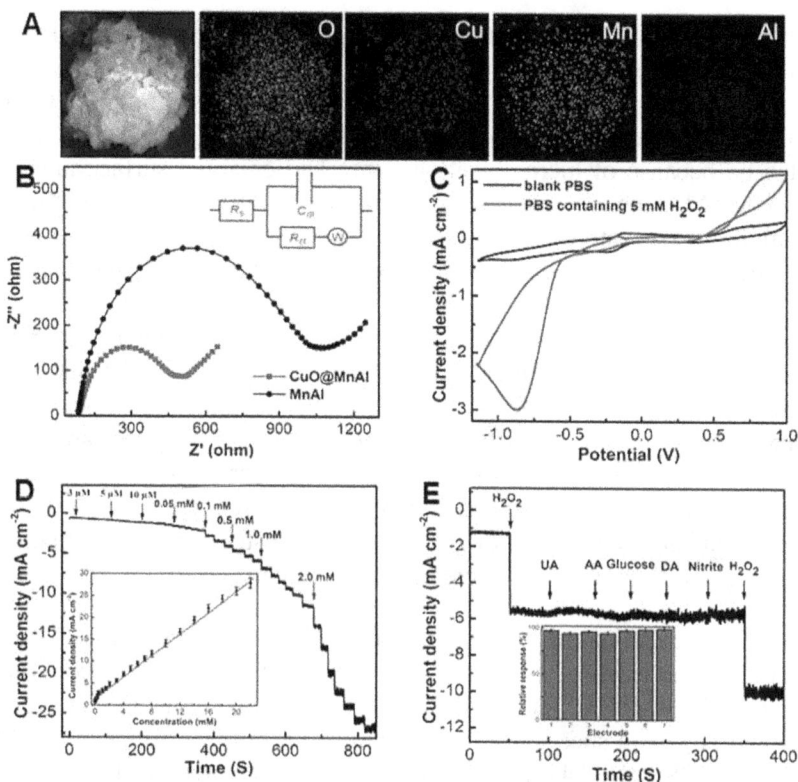

Fig. 5.3. (A) SEM-EDX mapping of CuO@MnAl NSs, (B) Nyquist plots of CuO@MnAl/GCE and MnAl/GCE in 0.1 M KCl containing 1.0 mM $K_3Fe(CN)_6$ and 1.0 mM $K_4Fe(CN)_6$. Frequency range: $0.1 \sim 10^5$ Hz. Inset is the equivalent circuit. (C) CV curves of CuO@MnAl/GCE in 0.1 M PBS in the absence (black line) and presence (red line) of 5 mM H_2O_2 (D) Amperometric i-t response of CuO@MnAl/GCE with successive additions of different H_2O_2 concentrations in N_2 saturated stirring 0.1 M PBS at an applied potential of - 0.85 V, Iset is the corresponding calibration curve of current density *vs.* H_2O_2 concentration. (E) Investigation of electroactive interferences on the response of 5 mM H_2O_2, Inset is the current response of seven different electrodes to 5 mM H_2O_2.

Cyclic voltammetry (CV) has been employed to assess the electrochemical and biosensing capabilities of Cu3-MnAl architectures. Fig. 5.3C depicts the CV curves of Cu3-MnAl LDH in 0.1 M deoxygenated phosphate buffer solution (PBS) with pH 7.4, in the presence (red line) and absence (black line) of 5 mM H_2O_2 at a scan rate of 100 mV/s. Cu3-MnAl modified electrode demonstrates a well-defined couple of redox peaks related to oxidation and reduction of H_2O_2, indicating the striking catalytic aptitude of the nanoclusters. A pair of minor peaks in the absence of H_2O_2 is consigned with the redox of Cu(I) to Cu(II) species. This is owing to the construction of p-n junction, may also be because of hierarchical structure and enlarged surface area of Cu3-MnAl nanocomposites that increase the absorbability of target analyte. In fact, appropriate insertion of porous n-type CuO species into p-type hydrotalcite layers, establishing p-n junction, enrich the number of charge carriers at breakdown voltage. The surface area of LDH precursors is going to increase by increasing the percent intercalation of Cu and then declines after conquering maximum of 185 m^2/g at intermediate composition of 3 mM (3.6 %) loading of copper. This improved surface area boosts up the active sites and stacking capability of H_2O_2 molecules on catalyst nanoclusters. Moreover, by sequential addition of H_2O_2 concentration in PBS solution, reduction peak current increased significantly, convincingly demonstrate the outstanding sensing ability of as-synthesized electrode nanomaterial.

Quantitative evaluation for H_2O_2 biosensing performance of Cu3-MnAl electrode was assessed by amperometric responses with a fixed potential of -0.85 V. Fig. 5.3D shows distinct steady state current-time feedback (i-t) curves upon successive inoculation of H_2O_2 in N_2 saturated 0.1 M stirring PBS (pH 7.4). A good stepwise enhancement in the current responses is noticed by augmenting H_2O_2 concentrations. Cu3-MnAl nanostruture biosensor achieved steady state current ~95 % within 3 s, representing quick transfer rate of electrons and swift diffusion of H_2O_2 on surface of electrode. Inset of Fig. 5.3E illustrates excellent linear detection range of 3 µM to 22 mM with correlation coefficient of 0.994 for Cu3-MnAl LDH based sensor that is suggestive of biosensor electrode's ability from nanomolar to millimolar detection scale for H_2O_2 concentration. The sensitivity of modified electrode has been measured with a detection limit down to 1 µM (signal/noise ratio of 3). Compared with other copper based nanomaterials such as DNA supported CuAl LDHs [30], grapheme wrapped Cu_2O nanocubes [25], CuO-Nafion [39] and CuS modified GC electrode [40], current LDHs nanomaterials are more attractive in having wide linear detection range and lowest detection limit. It could be noted that the present biosensor possess marvelous and inherent competency to detect H_2O_2 in terms of broad linear range, low detection limit and good sensitivity.

Moreover, the selectivity and anti-interfering capability of biosensor was evaluated amperometrically by similar response conditions with the addition of H_2O_2 and other electro-active analytes in PBS. Fig. 5.3E presents the amperometric influences of 5 mM each, uric acid (UA), ascorbic acid (AA), glucose, dopamine (DA), and nitrite in 0.1 M deoxygenated PBS solution that could not affect the detection of H_2O_2 under same environment. These findings show that the nonenzymatic H_2O_2 sensor functioning at particular potential is free from interfering electro-active species, because of relatively higher negative operating potential (-0.85 V) that limits the oxidation of foreigner interfering type of substances. To further inspect reproducibility and repeatedly of Cu3-MnAl electrode, amperometric responses of seven different electrodes were recorded with relative standard deviation (RSD) value not more than 5 % in 0.1 M PBS solution consisting 5 mM H_2O_2. In addition, biosensor has reproducible current values with (RSD) below 4 % for 5 mM H_2O_2 on seven consecutive assays (Fig. 5.3E inset).

5.7. Detection of H_2O_2 Triggered by Live Cancer Cells

Given to the optimal performance in the terms of incredible sensitivity, selectivity, broad linear range and reproducibility, CuO-MnAl based biosensor is reliable to use as tracker for *in-vitro* H_2O_2 secretion from live cells. Secretion of H_2O_2 was triggered by the inoculation of N-formyl-methionyl-leucyl-phenylalanine (fMLP), which resemble with metabolic oxidative route likely to be encountered *in-vivo* [41]. Fig. 5.4 depicts immediate current responses led by the addition of 10 μM (0.1 mM) fMLP to epithelial normal cells HBL-100, breast cancer cells MCF-7 and glioma brain cancer cells U87. No current signal was indicated by control well without cells suggesting the released of H_2O_2 by living cells. Furthermore, we can differentiate all these cell lines by current signal responses. The current response variation for non- tumorigenic epithelial HBL-100 cells is 0.92 μA; however the current changes for tumorigenic breast cancer cells MCF-7 and glioma brain cancer cells U87 are 2.41 μA and 4.97 μA respectively, which confirm the fast proliferation of cancerous cells to generate more H_2O_2. The aforementioned results are analogous with the evidences that tumorigenic cells proliferate quickly due to fast oxygen metabolism. This study reveals the reliable and robust application of CuO-MnAl LDHs as sensitive probe for quantitative determination of intracellular H_2O_2 efflux to segregate tumorigenic cells from normal ones.

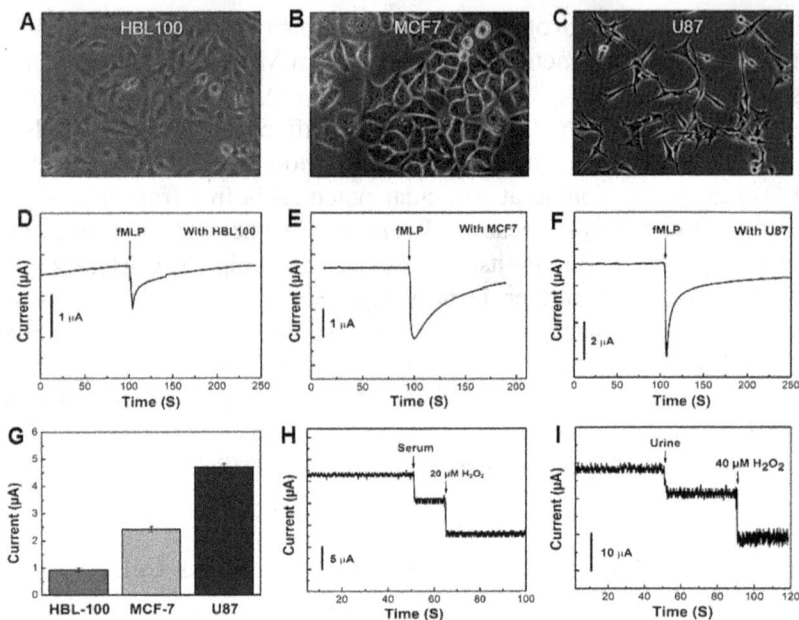

Fig. 5.4. Bright-field images of (A) HBL100 cells, (B) MCF7 cells and (C) U87 cells used in real-time tracking of H_2O_2. Amperometric responses of CuO@MnAl NSs/GCE in 0.1 M PBS (pH 7.4) upon the inoculation of 10 μL fMLP (0.1 mM) in the presence of (D) HBL100 cells, (E) MCF7 cells, and (F) U87 cells. (G) Increased current response of different live cells by adding 10 μL of fMLP (0.1 mM). Amperometric response of CuO@MnAl NSs upon an aliquot addition of (H) serum and (I) urine followed by successive addition of 20 μM and 40 μM H_2O_2 respectively in 0.1 M stirring PBS (pH 7.4).

Another LDHs based material such as Mn5-MgAl nanohybrids (SEM image of Fig. 5.5A) has been investigated for the sensitive detection of H_2O_2 from live cancer cells. Fig. 5.5 shows that a significant cathodic current was observed upon the inoculation of 10 μM fMLP for live HeLa cells, which went on declining gradually and reaches a plateau at 2 μA after 40 S. Furthermore, after the addition of catalase, a selective scavenger of H_2O_2, the current decreased progressively and went approximately to the background level. No any signal response was turned out by the injection of fMLP into control wells without HeLa live cancer cells. These consequences substantially declare that Mn5-MgAl based biosensor establishes a susceptible, reliable and robust method for quantitative *in vitro* and *in vivo* determination of H_2O_2 flux triggered by live cells.

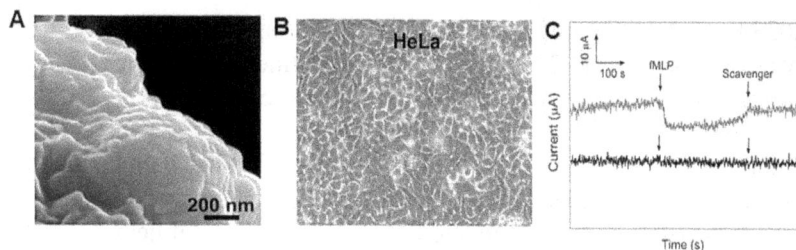

Fig. 5.5. (A) SEM images of synthesized Mn5-MgAl, (B) Bright field microscopy image of live HeLa cells. (C) Amperometric response of Mn5-MgAl GCE in 0.1 M PBS (pH 7.4) with the addition of 10 μM fMLP in the absence (upper) and presence (lower) of HeLa cells.

Moreover, core-shell Fe_3O_4@CuAl NSs LHDs [42] have been evaluated as biosensor in electrochemical sensing of H_2O_2 concentrations. Two different living cancer cell lines, *i.e.,* mouse hepatoma cell line H-22 and human hepatocellular carcinoma cell line BEL-7402 are chosen for real-time monitoring of H_2O_2 flux after being stimulated. And the bright-field images of these living cells are shown in Figs. 5.6A and 5.6D. BEL-7402 cell line is the fifth common malignancy, owing to notoriously high chemoresistive and scarcity of effective therapy, it is ranked as third public issue of cancer related death [43]. The mouse liver cancer cell line H-22 is one of the common malignancies, at most five years of survival after its diagnosis, threatening seriously global public health with high mortality and mostly diagnosed after getting an advanced stage [44]. The only way to improve resectability as well as survival rate and get rid of these lethal attacks is their early diagnosis. The endogenous H_2O_2 released was triggered by the addition of 0.1 mM fMLP. Upon the addition of 10 μL fMLP to the test wells, the significant reduction current densities of 0.121 mA cm^{-2} and 0.0891 mA cm^{-2} were observed in the presence of BEL-7402 and H-22 cells, respectively, as presented by Figs. 5.6B and 5.6E. Furthermore, current densities gradually decreased to a background level upon addition of catalase, indicating that released H_2O_2 is decomposed or diffused away from electrode in electrolytic solution. The control experiment without cells did not generate any detectable current signal, further suggesting that the triggered H_2O_2 efflux was due to stimulation of cells by fMLP (Fig. 5.6G). Because of enhanced plasma chemistry, the therapeutic efficacy of plasma activated medium (PAM) has been evaluated by detecting extracellular H_2O_2 released upon being stimulated by adding 10 μL of fMLP with proposed biosensor after plasma treatment. The significant decrease in current densities of

0.0334 mA cm^{-2}, 0.0088 mA cm^{-2} and 0.0196 mA cm^{-2}, 0.0068 mA cm^{-2} was observed for H-22 and BEL-7402 cell lines after being irradiated with plasma plumes for 1 min and 2 min, respectively (Figs. 5.6C and 5.6F), which decreased 72.39 %, 92.15 % and 78 %, 91 % in comparison to that without treatment (Fig. 5.6H). All these findings demonstrate the potential use of proposed biosensor for cancer diagnosis and enhanced therapeutic efficacy of plasma medicines for cancer treatments as well.

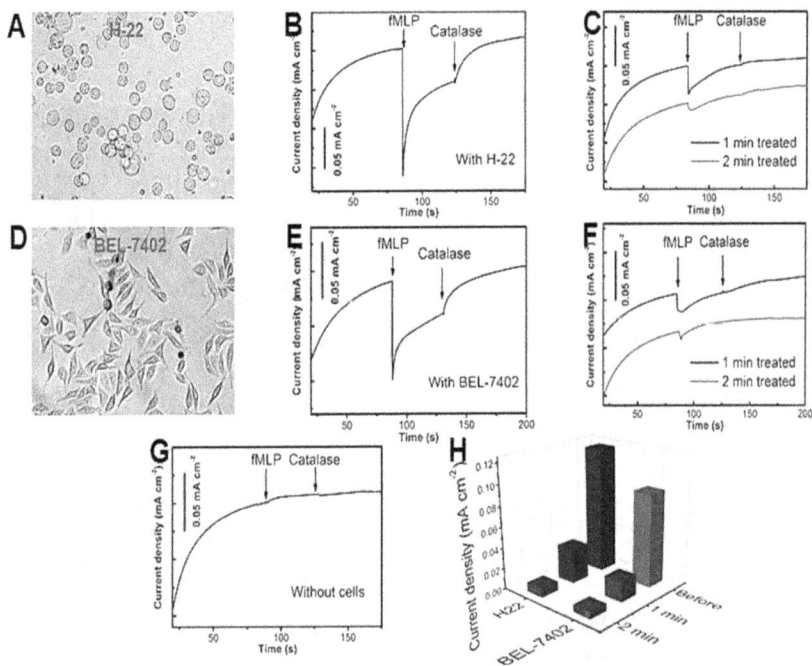

Fig. 5.6. Bright-field images of (A) H-22 cells, and (D) BEL-7402 cells. Amperometric responses of Fe$_3$O$_4$@CuAl/GCE upon the addition of 10 μL of fMLP in the tested wells with (B) H-22 cells, (E) BEL-7402 cells, and (G) without cells before plasma treatment, and (C) H-22 cells, and (F) BEL-7402 cells after 1 min (upper) and 2 min (lower) of plasma treatment, respectively. (H) Corresponding histogram of decreasing amperometric responses after plasma therapy.

5.8. Detection of H₂O₂ in Human Serum and Urine Samples

The endogenous H_2O_2 concentration was measured by amperometric response of as-synthesized biosensor upon the inoculation of biotic fluids. The aliquots of N_2 purged serum or urine analyte and 10 mL concentration of N_2 saturated 0.1 M PBS as electrolyte solution were placed in electrochemical cell with subsequent spiking of H_2O_2 concentrations and analyzed by Cu3-MnAl electrode as depicted by Figs. 5.4H and 5.4I. From current response vs concentration curve, the quantity of ubiquitous H_2O_2 in real samples of serum and urine was measured by using regression equation and found to be 14 μM and 61.5 μM respectively.

The TEM image of as-synthesize core-shell $Fe_3O_4@CuAl$ NSs LHDs material has shown in Fig. 5.7A. The endogenous amount of H_2O_2 from biofluids has been measured by proposed $Fe_3O_4@CuAl/GCE$ biosensor as presented by Fig. 5.7.

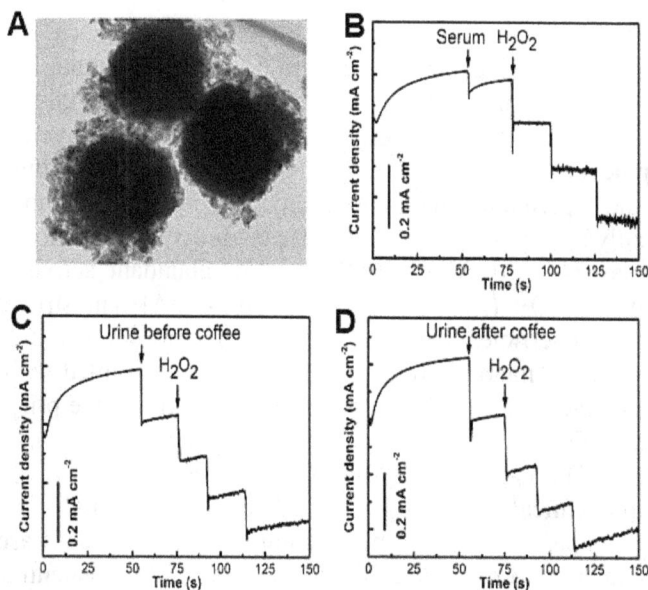

Fig. 5.7. (A) TEM image $Fe_3O_4@CuAl$ NSs. Amperometric responses of $Fe_3O_4@CuAl/GCE$ upon an aliquot addition of (B) serum, and (C) and (D) urine prior and following to the intake of coffee by successive addition of 20 μM and 30 μM H_2O_2, respectively.

The aliquots of serum and urine samples were inoculated into 15 mL of 0.1 M stirring PBS (pH 7.4) with subsequently spiking of some standard H_2O_2 concentrations. The ubiquitous amount of H_2O_2 excreted was determined by using regression equation and found the results for serum 12 μM, urine before 35.4 μM and after coffee intake 51.1 μM respectively. Considerable rise in urinary H_2O_2 level one hour after the intake of coffee has been noticed. Coffee beans are rich in hydroxy hydroquinone and polyphenols like chlorogenic acid, which is one possible cause of upsurge in urinary H_2O_2 concentration subsequently after coffee intake [45]. It can be concluded that these biological molecules are first absorbed into the body and then autoxidize upon exposure to oxygen, generating H_2O_2 after being excreted through urine. These results substantially demonstrate the reliable application of Fe_3O_4@CuAl NSs for cancer detection along with determination of ubiquitous H_2O_2 in biotic fluids.

5.9. Conclusions

In summary, we have developed various LDH nanomaterials, some inherent to p-n junction diodes by simple co-precipitation and hydrothermal techniques. The LDHs architectures ensure several distinct topographies such as: (i) Using MnAl LDHs with numerous charge carriers and superb intercalation mechanics for entrapping nanoparticles, significantly enhancing percentage mass loading of CuO nanospheres in the host layers of MnAl LDHs that endow abundant active sites for reduction of H_2O_2; (ii) Fabrication of CuO-MnAl nanostructures by simple, robust, efficient and facile co-precipitation and hydrothermal approach; (iii) In nanospheres, n-type porous CuO coupled with P-type semiconductive MnAl LDHs having intrinsic features like p-n junction serve as efficient electrocatalysis, surging the number of charge transporters. The as-prepared LDHs based nanomaterials have been used as electrochemical biosensors for detection of small metabolic molecules, in particular when they are implemented in the ultrasensitive *in-vitro* detection of emerging biomarker, i.e., H_2O_2 concentrations in living cells for early diagnosis of cancer. We envision that our structurally integrated strategy for metal oxides intercalated LDHs nano-architectures will provide new avenue to design afterward generation of bionanoelectronics and miniaturized biosensors as sensitive cancer detection probe.

References

[1]. S. K. Maji, S. Sreejith, A.K. Mandal, X. Ma, Y. Zhao, Immobilizing gold nanoparticles in mesoporous silica covered reduced graphene oxide: A hybrid material for cancer cell detection through hydrogen peroxide sensing, *ACS Applied Materials and Interfaces*, Vol. 2014, pp. 13648-13656.

[2]. S. Dong, J. Xi, Y. Wu, H. Liu, C. Fu, H. Liu, F. Xiao, High loading MnO$_2$ nanowires on graphene paper: facile electrochemical synthesis and use as flexible electrode for tracking hydrogen peroxide secretion in live cells, *Analytica Chimica Acta*, Vol. 853, 2015, pp. 200-206.

[3]. J. Bai, X. Jiang, A facile one-pot synthesis of copper sulfide-decorated reduced graphene oxide composites for enhanced detecting of H$_2$O$_2$ in biological environments, *Analytical Chemistry*, Vol. 85, 2013, pp. 8095-8101.

[4]. J. Wang, L. Cui, H. Yin, J. Dong, S. Ai, Determination of hydrogen peroxide based on calcined layered double hydroxide-modified glassy carbon electrode in flavored beverages, *Journal of Solid State Electrochemistry*, Vol. 16, 2011, pp. 1545-1550.

[5]. Y. Tian, F. Wang, Y. Liu, F. Pang, X. Zhang, Green synthesis of silver nanoparticles on nitrogen-doped graphene for hydrogen peroxide detection, *Electrochimica Acta*, Vol. 146, 2014, pp. 646-653.

[6]. F. Ai, H. Chen, S. H. Zhang, S. Y. Liu, F. Wei, X. Y. Dong, J. K. Cheng, W.H. Huang, Real-time monitoring of oxidative burst from single plant protoplasts using microelectrochemical sensors modified by platinum nanoparticles, *Analytical Chemistry*, Vol. 81, 2009, pp. 8453-8458.

[7]. Y. Zhang, X. Bai, X. Wang, K.K. Shiu, Y. Zhu, H. Jiang, Highly sensitive graphene-Pt nanocomposites amperometric biosensor and its application in living cell H$_2$O$_2$ detection, *Analytical Chemistry*, Vol. 86, 2014, pp. 9459-9465.

[8]. L. Zhu, Y. Zhang, P. Xu, W. Wen, X. Li, J. Xu, PtW/MoS2 hybrid nanocomposite for electrochemical sensing of H$_2$O$_2$ released from living cells, *Biosensors & Bioelectronics*, Vol. 80, 2016, pp. 601-606.

[9]. R. Weinstain, E. N. Savariar, C. N. Felsen, R. Y. Tsien, In vivo targeting of hydrogen peroxide by activatable cell-penetrating peptides, *Journal of the American Chemical Society*, Vol. 136, 2014, pp. 874-877.

[10]. H. Zhu, A. Sigdel, S. Zhang, D. Su, Z. Xi, Q. Li, S. Sun, Core/shell Au/MnO nanoparticles prepared through controlled oxidation of Aumn as an electrocatalyst for sensitive H$_2$O$_2$ detection, *Angewandte Chemie*, Vol. 126, 2014, pp. 12716-12720.

[11]. A. Gu, G. Wang, J. Gu, X. Zhang, B. Fang, An unusual H$_2$O$_2$ electrochemical sensor based on Ni(OH)$_2$ nanoplates grown on Cu substrate, *Electrochimica Acta*, Vol. 55, 2010, pp. 7182-7187.

[12]. R. Chinnasamy, G. Mohan Rao, R. T. Rajendra Kumar, Synthesis and electrocatalytic properties of manganese dioxide for non-enzymatic

hydrogen peroxide sensing, *Materials Science in Semiconductor Processing*, Vol. 31, 2015, pp. 709-714.

[13]. X.-M. Chen, Z.-X. Cai, Z.-Y. Huang, M. Oyama, Y.-Q. Jiang, X. Chen, Ultrafine palladium nanoparticles grown on graphene nanosheets for enhanced electrochemical sensing of hydrogen peroxide, *Electrochimica Acta*, Vol. 97, 2013, pp. 398-403.

[14]. I. A. Gorodetskaya, A. A. Gorodetsky, Analytical chemistry: Clamping down on cancer detection, *Nature Chemistry*, Vol. 7, 2015, pp. 541-542.

[15]. F. Qu, H. Sun, S. Zhang, J. You, M. Yang, Electrochemical sensing platform based on palladium modified ceria nanoparticles, *Electrochimica Acta*, Vol. 61, 2012, pp. 173-178.

[16]. Q. Zhao, R. Duan, J. Yuan, Y. Quan, H. Yang, M. Xi, A reusable localized surface plasmon resonance biosensor for quantitative detection of serum squamous cell carcinoma antigen in cervical cancer patients based on silver nanoparticles array, *International Journal of Nanomedicine*, Vol. 22, 2014, pp. 1097-1104.

[17]. V. V. Pérez, Synthesis of highly quenching fullerene derivatives for biosensor applications, MS Thesis, *Massachusetts Institute of Technology*, 2004.

[18]. X. Chi, D. Huang, Z. Zhao, Z. Zhou, Z. Yin, J. Gao, Nanoprobes for in vitro diagnostics of cancer and infectious diseases, *Biomaterials*, Vol. 33, 2012, pp. 189-206.

[19]. M. Asif, W. Haitao, D. Shuang, A. Aziz, G. Zhang, F. Xiao, H. Liu, Metal oxide intercalated layered double hydroxide nanosphere: With enhanced electrocatalyic activity towards H_2O_2 for biological applications, *Sensors and Actuators B: Chemical*, Vol. 239, 2017, pp. 243-252.

[20]. X. Sun, S. Guo, Y. Liu, S. Sun, Dumbbell-like Ptpd–Fe_3O_4 nanoparticles for enhanced electrochemical detection of H_2O_2, *Nano Letters*, Vol. 12, 2012, pp. 4859-4863.

[21]. Z. Xu, L. Yang, C. Xu, Pt@UiO-66 heterostructures for highly selective detection of hydrogen peroxide with an extended linear range, *Analytical Chemistry*, Vol. 87, 2015, pp. 3438-3444.

[22]. A. Kafi, A. Ahmadalinezhad, J. Wang, D .F. Thomas, A. Chen, Direct growth of nanoporous Au and its application in electrochemical biosensing, *Biosensors and Bioelectronics*, Vol. 25, 2010, pp. 2458-2463.

[23]. Y. Jin, H. Chen, M. Chen, N. Liu, Q. Li, Graphene-patched CNT/MnO_2 nanocomposite papers for the electrode of high-performance flexible asymmetric supercapacitors, *ACS Applied Materials and Interfaces*, Vol. 5, 2013, pp. 3408-3416.

[24]. F. Xiao, Y. Li, X. Zan, K. Liao, R. Xu, H. Duan, Growth of metal-metal oxide nanostructures on freestanding graphene paper for flexible biosensors, *Advanced Functional Materials*, Vol. 22, 2012, pp. 2487-2494.

[25]. M. Liu, R. Liu, W. Chen, Graphene wrapped Cu_2O nanocubes: Non-enzymatic electrochemical sensors for the detection of glucose and hydrogen peroxide with enhanced stability, *Biosensors & Bioelectronics*, Vol. 45, 2013, pp. 206-212.

[26]. S. L. Chou, J. Z. Wang, M. Choucair, H. K. Liu, J. A. Stride, S. X. Dou, Enhanced reversible lithium storage in a nanosize silicon/graphene composite, *Electrochemistry Communications*, Vol. 12, 2010, pp. 303-306.

[27]. Q. Zhang, S. Wu, L. Zhang, J. Lu, F. Verproot, Y. Liu, Z. Xing, J. Li, X. M. Song, Fabrication of polymeric ionic liquid/graphene nanocomposite for glucose oxidase immobilization and direct electrochemistry, *Biosensors & Bioelectronics*, Vol. 26, 2011, pp. 2632-2637.

[28]. N. Ding, N. Yan, C. Ren, X. Chen, Colorimetric determination of melamine in dairy products by Fe_3O_4 magnetic nanoparticles H_2O_2 ABTS detection system, *Analytical Chemistry*, Vol. 82, 2010, pp. 5897-5899.

[29]. Z. Li, J. Lu, S. Li, S. Qin, Y. Qin, Orderly ultrathin films based on perylene/poly (N-vinyl carbazole) assembled with layered double hydroxide nanosheets: 2D fluorescence resonance energy transfer and reversible fluorescence response for volatile organic compounds, *Advanced Materials*, Vol. 24, 2012, pp. 6053-6057.

[30]. L. Chen, K. Sun, P. Li, X. Fan, J. Sun, S. Ai, DNA-enhanced peroxidase-like activity of layered double hydroxide nanosheets and applications in H_2O_2 and glucose sensing, *Nanoscale*, Vol. 5, 2013, pp. 10982-10988.

[31]. A. Jawad, Y. Li, X. Lu, Z. Chen, W. Liu, G. Yin, Controlled leaching with prolonged activity for Co-LDH supported catalyst during treatment of organic dyes using bicarbonate activation of hydrogen peroxide, *Journal of Hazardous Materials*, Vol. 289, 2015, pp. 165-173.

[32]. S. Yuan, D. Peng, X. Hu, J. Gong, Bifunctional sensor of pentachlorophenol and copper ions based on nanostructured hybrid films of humic acid and exfoliated layered double hydroxide via a facile layer-by-layer assembly, *Analytica Chimica Acta*, Vol. 785, 2013, pp. 34-42.

[33]. X. Han, K. Fang, J. Zhou, L. Zhao, Y. Sun, Synthesis of higher alcohols over highly dispersed cu-fe based catalysts derived from layered double hydroxides, *Journal of Colloid and Interface Science*, Vol. 470, 2016, pp. 162-171.

[34]. D. Shan, S. Cosnier, C. Mousty, Layered double hydroxides: An attractive material for electrochemical biosensor design, *Analytical Chemistry*, Vol. 75, 2003, pp. 3872-3879.

[35]. Y. Wang, W. Peng, L. Liu, F. Gao, M. Li, The electrochemical determination of L-cysteine at a Ce-doped Mg-Al layered double hydroxide modified glassy carbon electrode, *Electrochimica Acta*, Vol. 70, 2012, pp. 193-198.

[36]. M. Asif, A. Aziz, A. Q. Dao, A. Hakeem, H. Wang, S. Dong, G. Zhang, F. Xiao, H. Liu, Real-time tracking of hydrogen peroxide secreted by live cells using MnO_2 nanoparticles intercalated layered doubled hydroxide nanohybrids, *Analytica Chimica Acta*, Vol. 898, 2015, pp. 34-41.

[37]. F. Xiao, J. Song, H. Gao, X. Zan, R. Xu, H. Duan, Coating graphene paper with 2D-assembly of electrocatalytic nanoparticles: A modular approach

toward high-performance flexible electrodes, *ACS Nano*, Vol. 6, 2011, pp. 100-110.

[38]. X. Wen, Z. Yang, X. Xie, Z. Feng, J. Huang, The effects of element cu on the electrochemical performances of zinc-aluminum-hydrotalcites in zinc/nickel secondary battery, *Electrochimica Acta*, Vol. 180, 2015, pp. 451-459.

[39]. P. Gao, D. Liu, Facile synthesis of copper oxide nanostructures and their application in non-enzymatic hydrogen peroxide sensing, *Sensors and Actuators B: Chemical*, Vol. 208, 2015, pp. 346-354.

[40]. A. K. Dutta, S. Das, P. K. Samanta, S. Roy, B. Adhikary, P. Biswas, Non-enzymatic amperometric sensing of hydrogen peroxide at a CuS modified electrode for the determination of urine H_2O_2, *Electrochimica Acta*, Vol. 144, 2014, pp. 282-287.

[41]. J. Xi, Y. Zhang, N. Wang, L. Wang, Z. Zhang, F. Xiao, S. Wang, Ultrafine Pd nanoparticles encapsulated in microporous Co_3O_4 hollow nanospheres for in situ molecular detection of living cells, *ACS Applied Material and Interfaces*, Vol. 7, 2015, pp. 5583-5590.

[42]. M. Asif, H. Liu, A. Aziz, H. Wang, Z. Wang, M. Ajmal, F. Xiao, Core-shell iron oxide-layered double hydroxide: High electrochemical sensing performance of H_2O_2 biomarker in live cancer cells with plasma therapeutics, *Biosensors & Bioelectronics*, Vol. 97, 2017, pp. 352-359.

[43]. Z. Zhang, C. Zhang, Y. Ding, Q. Zhao, L. Yang, J. Ling, L. Liu, H. Ji, Y. Zhang, The activation of p38 and JNK by ROS, Contribute to OLO-2-mediated intrinsic apoptosis in human hepatocellular carcinoma cells, *Food Chemistry Toxicology*, Vol. 63, 2014, pp. 38-47.

[44]. J. Yang, X. Li, Y. Xue, N. Wang, W. Liu, Anti-hepatoma activity and mechanism of corn silk polysaccharides in H22 tumor-bearing mice, *Internation Journal of Biological Macromolicules*, Vol. 64, 2014, pp. 276-280.

[45]. S. Chatterjee, A. Chen, Functionalization of carbon buckypaper for the sensitive determination of hydrogen peroxide in human urine, *Biosensors & Bioelectronics*, Vol. 35, 2012, pp. 302-307.

Chapter 6

Highly Reliable Metallization on Polymer and Their Fundamental Characteristics Toward Wearable Devices Applications

Wan-Ting Chiu, Byung-Hoon Woo, Mitsuo Sano, Tso-Fu Mark Chang, Chun-Yi Chen, Tomoko Hashimoto, Hiromichi Kurosu and Masato Sone

6.1. Introduction

The worldwide wearable device market is foreseen to swell to 578 million USD by 2019 projecting the trend and blossom of the promising technology [1]. Wearable devices can be practiced to various applications such as pacemakers tracking the heartbeat and sportswear recording the body status for advance of the sports science. Notwithstanding several wearable devices have been proposed and investigated, many of the apparatus are fabricated with inflexible and rigid substrates. For an active user such as a jogger, even a slight irritation might be offensive. Therefore, a conductive layer coated on flexible fabrics for the wearable devices is now urgently demanded for development of the next–generation technology [2]. Combining flexible and conductive materials thus becomes the greatest challenge. There are many ways to integrate two distinct materials together, such as sputtering, ink printing, dipping process [3], and electroless plating [4]. Among the processes, electroless plating, a prevalent technique, surpasses other processes due to its low cost and moderate deposition rate [5]. Electroless plating is also known as a chemical plating that can be operated without external electrical power and is especially advantageous when the substrate is not electrical conductive and having

Wan-Ting Chiu
Institute of Innovative Research, Tokyo Institute of Technology,
CREST, Science and Technology Agency, Japan

sophisticated shape. Furthermore, conformal deposition are critical requirements for the composite materials. Electroless plating is a powerful technique, which can give conformal deposition [4]. There are three major steps in electroless plating, the first step is a pretreatment step to clean and roughen the substrate. The second step is to embed activate nucleation sites onto the substrate by a catalyzation process, and metallization is the last step to metallized the substrate by depositing metal coatings on the substrate.

An advanced catalyzation technique facilitating supercritical CO_2 (sc-CO_2) was applied to overcome the most challenging problem in the catalyzation step of electroless plating [6]. In the conventional (CONV) catalyzation, catalyst-contained corrosive solution is used to embed the catalyst, however, the substrate structure is often damaged by the corrosive solution during the process. Due to the exceptional self-diffusivity [7], low surface tension [8], and affinity to non-polar materials of sc-CO_2 [9], the catalyst can be inlaid into the substrate by using sc-CO_2 while remains the substrate structure persistent [10]. Moreover, due to the introduction of sc-CO_2, the electroless plating steps could be simplified into two steps by means of withdrawing the pretreatment step, because the catalyzation can be completed without the roughening step. Therefore, an electroless plating method promoted by sc-CO_2 has been developed in effort to advance the plating characteristics [11]. On the other hand, in CONV metallization, reactions are usually accompanied with hydrogen evolutions, which could lead to defects in the metallization layer [11]. Sc-CO_2 is introduced into the metallization step to enhance the metallization properties. A surfactant is introduced into the aqueous metallization bath to emulsify the aqueous bath and form surfactant/sc-CO_2 micelles [11]. With the help of the micelles, it can carry the hydrogen bubbles away from the substrate due to its non-polarity and high self-diffusivity. The metallization layer turns out to be compact and smooth by elimination of the hydrogen bubbles. Sc-CO_2 was therefore introduced catalyzation and metallization steps to enhance the plating characteristics.

In the substrate point of view, polyimide, Nylon 6,6 textile, and silk textile were chosen as the substrates for the interest of their biocompatibility, flexibility, stitchability, high strength, and acting as a common material of clothing. In the metallization point of view, Ni has been practiced to the wearable devices due to the controllable deposition, low cost, and high corrosion resistance [12]. On the other hand, due to the high biocompatibility and the passive property, Pt is often applied as the essential components of medical devices, especially implantable devices [13]. The electroless plating carried out by bis (2,4-pentandionato)-palladium Pd(acac)$_2$ and bis (2,4-pentandionato)-platinum Pt(acac)$_2$ catalyzation followed by Ni-P and Pt metallization, is now urgently demanded for the wearable and implantable medical devices.

For evaluating reliability of the composite materials, various evaluations were executed. Corrosion resistance is a critical property for the application due to the exposure to liquids during the employment. Corrosion tests were carried out in two different solutions to evaluate the corrosion resistances of the meta-polymer composite materials. Furthermore, a critical issue concerning biocompatibility assessment mentioned in the previous paragraph must be addressed for the final application in wearable and medical devises. According to J. C. Wataha [14], allergy reactions would be triggered when there is metal ion released from the environment and enters human body. The metal ion released from the composite becomes a criteria for evaluating the biocompatibility. For this reason, metal ion-releasing rate was examined by immersing the composite in a simulated body fluid (SBF) (r-type) [15] for three months to assess the biocompatibility. SBF is composed of various ions, which is similar to human body fluid. In addition, electrical conductivity is an essential property for applications in electronic devices. Hence electrical resistance of the composite was evaluated by an in-line four-point probe measurement (Mitsubishi Chemical Analytech Co., Ltd., MCP-T370, Japan). On the other hand, wearable devices are subjected to the external disturbance frequently while being used, assessments for the adhesive reliability thus is also a fundamental prerequisite. An adhesive test to evaluate the adhesive firmness between the metal layer and the polymer substrate was conducted in this study.

6.2. Nickel-Phosphorus Metallization

6.2.1. Nickel-Phosphorus Metallization on Polyimide

6.2.1.1. Sc-CO$_2$ Catalyzation and CONV Metallization

A square sheet of polyimide Kapton$^©$ (1.0×2.0 cm^2; thickness, 130 μm) was used as the substrate. CO$_2$ with a minimum purity of 99.9 % was purchased from Nippon Tansan Gas Co., Ltd. The Ni-P metallization solution had a chemical composition of nickel chloride (9 wt.%), sodium hypophosphite (12 wt.%), complexing agent (12 wt.%), and ion-exchanged water (67 wt.%) (Okuno Chemical Industries Co., Ltd.). The high-pressure experimental apparatus was made by Japan Spectra Company (Fig. 6.1). The Pd nucleus size and nuclei distribution on the substrate were observed with an FE-SEM. An XRD analysis was performed on the catalyzed polyimide surface with an X-ray diffractometer instrument to ascertain the features of the Pd nuclei. The X-ray was generated by a Cu target operated at 40 kV and 40 mA. In CONV catalyzation, the polyimide substrate was immersed in a mixed $PdCl_2/HCl$ solution (0.04 % $PdCl_2$ and 18 % HCl) for 20 min and washed with HCl. After immersion of the polyimide substrate in sc-CO_2 for catalyzation, $Pd(acac)_2$ (concentration: 1.219×10^{-6} mol/L) was deposited onto the substrate in sc-CO_2 for 20 min. The reaction temperature and pressure were 80 °C and 15 MPa, respectively. Visual point counting was selected to estimate the amount of Pd nuclei in the sc-CO_2 catalyzation procedure.

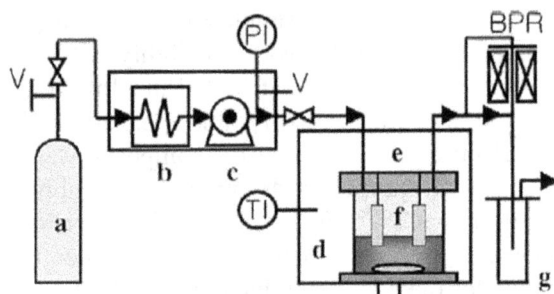

Fig. 6.1. (a) CO_2 gas tank, (b) liquidization unit, (c) high-pressure pump, (d) thermal bath, (e) reaction cell, (f) stirrer + substrate (×2) + Ni-P metallization sol. + sc-CO_2 + surfactant, (g) trap, BPR: back pressure, PI: pressure sensor, TI: thermometer, V: valve.

The photographs in Fig. 6.2 show the un-metallized polyimide substrate and the substrates metallized with Ni-P thin film by different methods. Fig. 6.2a shows a pure polyimide substrate. Fig. 6.2b shows a Ni-P thin film partially metallized for 5 min over the substrate after catalyzing the substrate by the CONV method in a mixed $PdCl_2/HCl$ solution at room temperature for 20 min. Fig. 6.2c shows totally metallized Ni-P thin film

on polyimide substrate after sc-CO_2 catalyzation with Pd
organo-complex at a reaction temperature and pressure of 357 K and
15 MPa, respectively. Noting these results, we concluded that sc-CO_2
had a good effect for total Ni-P coverage on the polymer substrate. This
difference was originated from catalyzation procedure. Thus we
observed catalyzed polyimide with SEM.

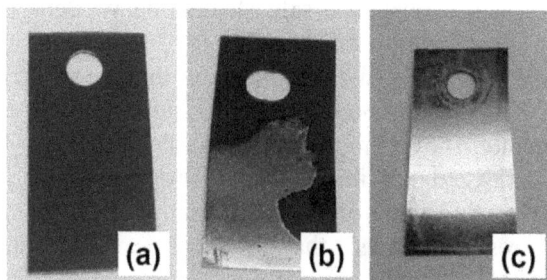

Fig. 6.2. Photographic images of polyimide (a), Ni-P metallization for 5 min
on CONV catalyzation for 20 min (b) and catalyzation
in sc-CO_2 for 20 min (c).

Figs. 6.3a and b shows SEM images of catalyzed substrates immersed
for 20 min. The Pd catalysts localized and aggregated on the CONV
catalyzed substrate (Fig. 6.3a) indicate that Pd grew on Pd nuclei initially
deposited on the polyimide. This is why several large nuclei are
observed. In contrast, the catalysts are well deposited and uniformly
distributed in the sc-CO_2-catalyzed substrate immersed for 20 min
(Fig. 6.3b). This difference follows from the compatibility of sc-CO_2
with hydrophobic polyimide. Sc-CO_2 diffusibility can also induce the
Pd-complex to disperse over all of the surface-treated polyimide, thus
facilitating uniform catalyzation growth. Polyimide substrates catalyzed
by both methods were metallized with Ni-P metallization solution with
a 5 min metallization time. Figs. 6.3c and d show optical microscopic
(OM) (Digital Microscope VHX-500, KEYENCE. CO., Ltd) images on
a covered Ni-P thin film catalyzed by the CONV method and by sc-CO_2.
The deposited Ni-P thin film treated by CONV catalyzation has profuse
cracking. In contrast, the Ni-P thin films treated by sc-CO_2 catalyzation
are free of cracks and are deposited over the entire surface. The cracks
in the former film are thought to result from the low catalyst density and
poor uniformity of the catalysts on the polyimide substrate. The low

169

density and ununiformity of the Pd on the polyimide substrate reduce the rate of Ni-P film coverage. According to these results, the catalyzation influences the Ni-P coverage on a macro-size scale, as well as the micro-size defects in the metallized Ni-P film. In the next, we describe the catalyst in the sc-CO_2 catalyzation on polyimide. It is essential to clarify Pd catalysts after the catalyzation procedure for Ni-P metallization to form a thin film. In EDX result, aggregated white particles with a diameter of about 50 μm in Fig. 6.3a indicates Cl and weak Pd. $PdCl_2$ was deposited as salt. Meanwhile, we detect Pd peaks on the polyimide surface catalyzed in sc-CO_2 for 20 min, and other peaks of C and O in white particles. We can thus infer that the white particles are Pd(acac)$_2$. In XRD results both polyimide and polyimide catalyzed by the CONV method shows amorphous peaks from the polyimide. In the results of the catalyzed polyimide substrate in sc-CO_2, we observe strong Pd(acac)$_2$ peaks. And there are no palladium and no palladium oxide peak. This result shows that Pd(acac)$_2$ were deposited as organo-complex.

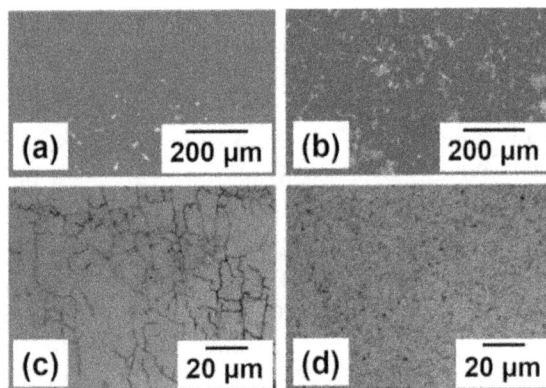

Fig. 6.3. SEM images of catalyzed substrate with CONV catalyzation for 20 min (a), catalyzation in sc-CO_2 for 20 min (b), OM images of metallized Ni-P film for 5 min after CONV catalyzation for 20 min (c), and sc-CO_2 catalyzation for 20 min (d).

For knowing catalyst growth, we performed catalyzation with various time lengths (5, 10, 15 and 20 min). In CONV catalyzation, the number of white particles increased with time, and their deposition was irregular. The catalyzation solution used for the CONV method was incompatible with the hydrophobic surface of the polyimide substrate. This hindered the nucleation of the Pd on the polyimide substrate, leading to the localized pattern of distribution. Irregular and localized deposited catalyst nuclei affect metal film coverage. The metallized surfaces catalyzed for 5, 10, and 15 min show small island-like formations metallized with Ni-P. It was difficult to conduct metallization with immersion times of less than 15 min. On sample catalyzed for 20 min, the Ni-P film was partially deposited. Autocatalytic reduction reactions in solution generally need an induction time to attain total coverage of a polyimide substrate. In our result, more than 20 min was essential as the induction time in the CONV catalyzation reaction for total Ni-P coverage. Meanwhile, catalysts are well deposited and distributed in the film metallized within a short period of sc-CO_2 catalyzation. The amount and size of these catalysts increased as the catalyzation in the sc-CO_2 proceeded over time. The crystal growth of the catalyst commenced when the catalyzation time reached 10 min. We also observed the deposition of abundant catalysts on the substrate within a relatively short time. This was principally because the diffusion coefficient of sc-CO_2 media is 10 or 100 times higher than aqueous Pd activation. Sc-CO_2 diffusibility could also encourage the Pd-complex to disperse over all of the surface-treated polyimide, thereby facilitating uniform catalyzation growth. Because sc-CO_2 is a hydrophobic media, it has good wettability with the polymer substrate.

To precisely clarify nucleation and nuclei growth in catalyzation, we quantitatively analyzed the catalyzation nucleation in sc-CO_2 using an SEM image. Fig. 6.4 shows the areas over which catalyst particles occupy a polyimide substrate catalyzed in sc-CO_2 at different catalyzation times. As the Fig. 6.4 shows, the substrate area occupied by the catalyst increased as the reaction time increased. At catalyzation times from 1 to 3 min, the area occupied by the catalyst ranged from 1 % to 3 %. At a catalyzation time of 5 min, the area occupied by the catalyst increased to about 7 %. After 5 min of catalyzation, the area occupied by the catalyst on polyimide substrate increased quasi-linearly. The area of the substrate occupied by the catalyst rapidly increased from 3 to 5 min.

Fig. 6.5a show Ni-P coverage ratios at different catalyzation times, from 1 min to 5 min. Figs. 6.5b-d shows OM images of Ni-P thin film surface after sc-CO_2 catalyzation time from 1 to 5 min. The Ni-P thin film after catalyzation for 1-3 min formed an island structure (Figs. 6.5b and c) and then merged to form a continuous structure after catalyzation for 5 min

(Fig. 6.5d). Complete coverage was not attained with catalyzation from 1 to 3 min, whereas it was obtained with catalyzation for 5 min in sc-CO_2. Here we observe a dramatic change from 3 to 5 min, compared to the result shown in Fig. 6.4.

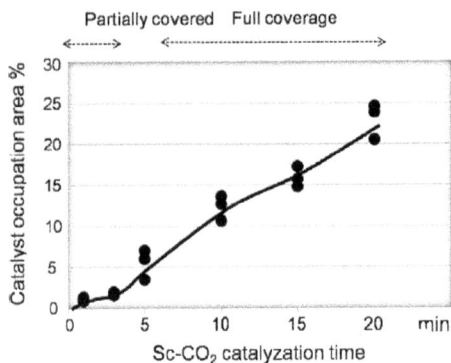

Fig. 6.4. Catalyst occupation area on polyimide substrate catalyzed in sc-CO_2 with various catalyzation time.

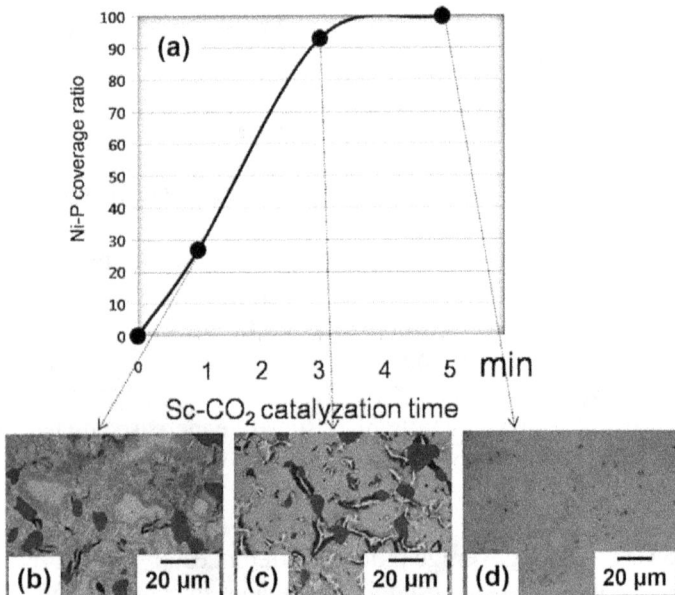

Fig. 6.5. Ni-P coverage percentage on different sc-CO_2 catalyzation time (a), OM images of Ni-P thin film surface with different sc-CO_2 catalyzation time (b) 1 min, (c) 3 min, and (d) 5 min.

This period is considered to be a critical point for full coverage. This result indicates that an area covered by critical catalyst is a necessary precondition for continuous coverage of the Ni-P thin film. Moreover, the catalyst sizes are also thought to influence the coverage. Though the areas occupied by the catalyzation changed only moderately with the change in the catalyzation time from 1 to 3 min, the coverage of Ni-P catalyzed for 3 min was about three times higher than that catalyzed 1 min. On this basis, we know that the increase of the catalyst size with the catalyzation time influences the metallization coverage. The catalyst size and the area of the polymer substrate occupied by the catalyst contribute strongly to the completely continuous coverage of the Ni-P film over the polymer substrate.

6.2.1.2. Effects of Sc-CO$_2$ Catalyzation in Metallization

A square sheet of polyimide Kapton$^©$ (1.0×2.0 cm^2; thickness, 130 μm) was used as the substrate. The substrate was washed with acetone and rinsed in ion exchanged water before each reaction. Grease was removed from the sample by successive dipping in a 10 wt.% solution of NaOH and a 10 wt.% solution of HCl, followed by rinsing in deionized water. The temperature variation of each run was confirmed to be less than 1.0 °C. The maximum working temperature and the maximum pressure were 150 °C and 50 MPa, respectively. The reaction chamber was a stainless steel 316 vessel with a volume of 50 ml, kept in a temperature-controlled air bath. A plastic-coated stainless steel 301 wire was fed through one of several holes in the chamber cap to connect and hang the polyimide substrate. A magnetic agitator with a cross-magnetic stirrer bar was placed within the reaction chamber, and the substrate was attached to the reactor with stainless steel 301 wires. An OM and FE-SEM were used to observe the surfaces of the metallized Ni-P films. The Pd nucleus size and nuclei distribution on the substrate were observed with an FE-SEM. Similar XRD analysis was performed to ascertain the features of the Pd nuclei. Detected peaks were analyzed with International Center for Diffraction Data base. Two types of catalyzation procedure were performed. After the pretreatment, the polyimide substrate was immersed in a CONV catalyzation solution (same as Section 6.2.1.1) for 20 min and washed with deionized water. After immersion of the polyimide substrate in sc-CO$_2$ for catalyzation, Pd(acac)$_2$ (concentration: 1.219×10^{-6} mol/L) was deposited onto the substrate in sc-CO$_2$ for 20 min. The reaction temperature and pressure were 80 °C and 15 MPa, respectively. Catalyzed polyimide substrate was immersed in a Ni-P metallization solution, which is the same as that in Section 6.2.1.1. CONV catalyzed polyimide substrate and catalyzed polyimide in sc-CO_2 were immersed in the Ni-P metallization solution. Metallization was conducted at 80 °C. Visual point counting was selected to estimate the amount of catalyst nuclei in the sc-CO_2 catalyzation procedure on polyimide substrate. The catalyst nucleus size and nuclei distribution on the substrate after cleaning with ethanol for 1 min were observed with a FE-SEM. Five different fields (216×216 µm) were selected for the analysis. The mesh was 27×31. To study the Ni-P surface coverage, a series of image analysis tools were applied to the OM images. We measured the Ni-P coverage with UTHSCSA image tool.

The photographs in Fig. 6.6 show the unmetallized polyimide substrate and the substrates metallized with Ni-P thin film by different methods. Fig. 6.6a shows a polyimide substrate. Fig. 6.6b shows a Ni-P thin film partially metallized over the substrate for 5 min in metallization solution followed by catalyzation by the CONV method in a mixed $PdCl_2$/HCl solution at room temperature for 20 min. Fig. 6.6c shows totally metallized Ni-P thin film for 5 min on polyimide substrate after sc-CO_2 catalyzation with Pd(acac)$_2$ at a reaction temperature and pressure of 80 °C and 15 MPa, respectively. According to these results, we concluded that sc-CO_2 catalyzation had a good effect for total Ni-P coverage on the polymer substrate comparing with CONV catalyzation method. The results also indicated that the sc-CO_2 catalyzation induced the Pd nuclei to deposit well and distribute uniformly on all parts of the polymer substrate for conformal and uniform Ni-P thin film. With CONV catalyzation, on the other hand, the catalysts tended to concentrate in certain areas, depleting the quantities available for coverage of other areas. This is why it is difficult, in CONV catalyzation, for palladium to catalyze uniformly on a polymer substrate.

Fig. 6.7 shows SEM images of catalyzed substrates immersed in CONV catalyzation (Fig. 6.7a) and sc-CO_2 catalyzation (Fig. 6.7b) for 20 min. There are localized and aggregated catalysts on the CONV catalyzed substrate. Moreover several large catalyst nuclei were observed. In contrast, the catalysts are well deposited and uniformly distributed in the sc-CO_2 catalyzed substrate immersed for 20 min. And catalysts were easily deposited on the polyimide in sc-CO_2. This difference seems to follow from the compatibility of sc-CO_2 with hydrophobic polyimide. Affinity between polymer and sc-CO_2 could induce Pd(acac)$_2$ to disperse over all of the surface treated polyimide, thus facilitating uniform catalyzation growth. The abundance of catalyst nuclei facilitated the deposition of the Ni-P thin film onto the polyimide substrate.

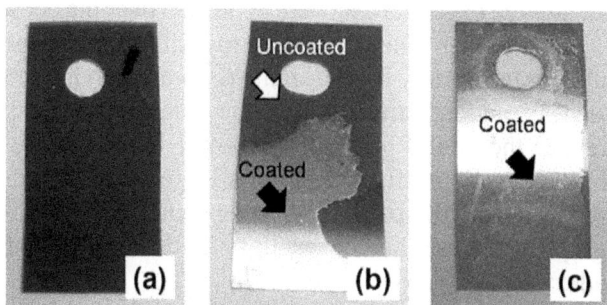

Fig. 6.6. Photographic images of polyimide (a), Ni-P metallization for 5 min on CONV catalyzation for 20 min (b) and catalyzation in sc-CO_2 for 20 min (c). A white arrow indicates uncoated area and black arrows indicate coated area.

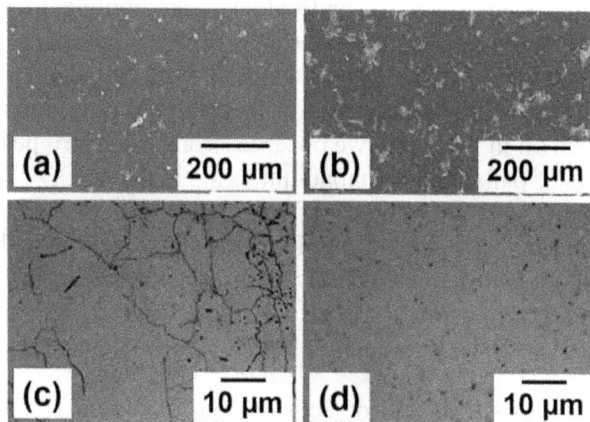

Fig. 6.7. SEM images of catalyzed substrate with CONV catalyzation for 20 min (a), catalyzation in sc-CO_2 for 20 min (b) and OM images of metallized Ni-P film for 5 min after CONV catalyzation for 20 min (c), and sc-CO_2 catalyzation for 20 min (d).

We observed on fabricated Ni-P thin film with OM after metallization reaction by both methods. Figs. 6.7c and d show OM images on a Ni-P thin film catalyzed by the CONV method and by sc-CO_2 respectively.

175

Fig. 6.7c shows only localized parts of the metallized Ni-P film on the substrate, which were partially covered by Ni-P as shown in Fig. 6.6b. These metallized Ni-P thin film treated by CONV catalyzation has lots of cracks. In contrast, the metallized Ni-P thin films treated by sc-CO$_2$ catalyzation are free of cracks and were deposited over the entire surface. The cracks in the former film are thought to result from the low catalyst density and poor uniformity of the catalysts on the polyimide substrate. The low density and ununiformity of the catalyst on the polyimide substrate reduce the rate of Ni-P film coverage. According to Figs. 6.6 and 6.7, the catalyzation influences the Ni-P coverage on a macro-size scale, as well as the micro-size defects in the metallized Ni-P film. In the next paragraph, we describe the catalyst in the sc-CO$_2$ catalyzation on polyimide.

In the current experiments, we used Pd(acac)$_2$ for the sc-CO$_2$ catalyzation. Thus it is essential to clarify Pd catalysts after the catalyzation procedure for Ni-P metallization. Fig. 6.8 shows the results of an XRD analysis performed to characterize the Pd catalysts with both CONV catalyzation method and sc-CO$_2$ catalyzation on the polyimide substrate. Fig. 6.8a shows the individual peaks coming from packing structure of Kapton. In Fig. 6.8b we observe only individual peaks of the polyimide catalyzed by the CONV method, with no peaks of the pure PdCl or PdO$_2$. Next, in the results for the catalyzed polyimide substrate in sc-CO$_2$ shown in Fig. 6.8c, we observe strong Pd(acac)$_2$ peaks. There are no peaks of palladium or palladium oxide in the sc-CO$_2$. This result indicates that the large amounts of acetylacetonate were deposited as organo-complex. Moreover, though we find no peaks of palladium or palladium oxide in the sc-CO$_2$ catalyzation, the Ni-P metal thin film was deposited after immersion in the metallization solution. This shows that the Pd(acac)$_2$ can serve as a catalyst for Ni-P metal deposition.

The XRD analysis failed to detect peaks of Pd and Pd oxide on the catalyzed polyimide obtained by the CONV method. There are possibilities that the pure palladium could be too scanty for XRD to detect on the substrate. Thus we decided to perform SEM and EDX analyses of the polyimide surface catalyzed by the CONV method for 20 min. Fig. 6.9a shows aggregated white particles with a diameter of about 30 μm. The EDX results of analysis on the particles show strong Cl peaks and weak palladium peaks. When a polymer substrate is immersed in a mixed solution of PdCl and HCl, the Pd catalyst can generally be deposited on a substrate. Thus, PdCl was deposited as salt. We also note that the Pd and Cl peaks have similar peak areas. In sc-CO$_2$

catalyzation for 20 min, observed white particles on the polyimide show Pd peaks and other peaks of C and O in white particles. We can thus infer that the white particles are $Pd(acac)_2$.

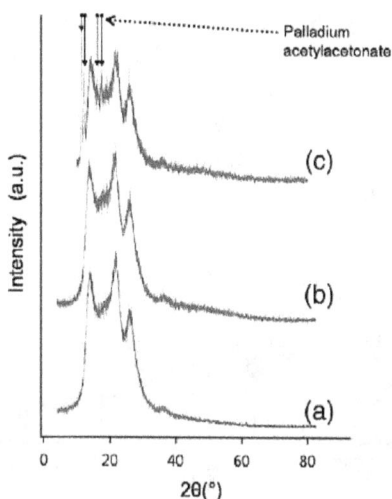

Fig. 6.8. XRD spectrum of polyimide substrate (a), catalyzed polyimide substrate with CONV catalyzation for 20 min (b) and sc-CO_2 catalyzation for 20 min (c).

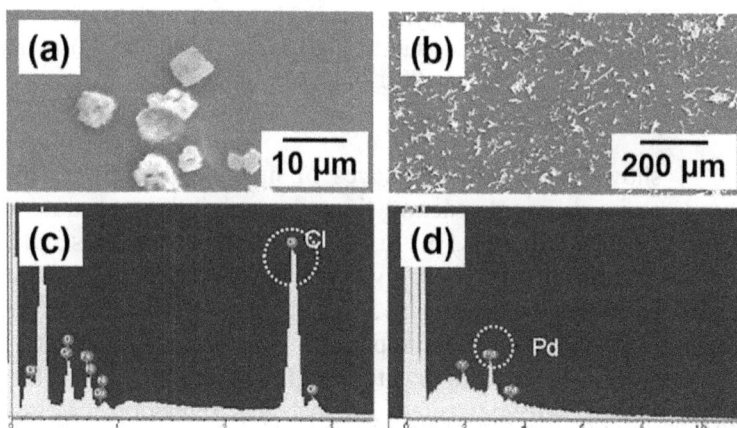

Fig. 6.9. SEM images (a), (b) and EDX results (c), (d) of catalyzed polymer substrate by CONV catalyzation for 20 min (a), (c) and sc-CO_2 catalyzation for 20 min (b), (d).

177

Fig. 6.10 shows SEM images of polyimide surfaces catalyzed for different catalyzation times. White particles can be observed in some parts of the catalyzed surfaces after 10 min. Cl and Pd atoms were detected within them. The amount of white particles a little increased with time, and their deposition was irregular. Thus, some large nuclei can be observed in Fig. 6.10d. It can be difficult to uniformly deposit catalysts on polymer substrate by the CONV method. The catalyzation solution used for the CONV method was incompatible with the hydrophobic surface of the polyimide substrate. This hindered the nucleation of the Pd on the polyimide substrate, leading to the localized pattern of distribution.

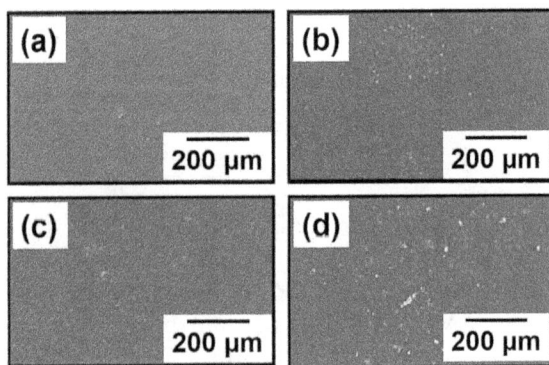

Fig. 6.10. SEM images of Pd catalyzed polyimide surface with CONV method for different times on polyimide substrate (a) 5 min, (b) 10 min, (c) 15 min, and (d) 20 min.

Fig. 6.11 shows photographic images of metallized film by CONV catalyzation for different reaction times on polyimide substrate. The metallized surfaces in Figs. 6.11a, b, and c show small, island like formations metallized with Ni-P. It was difficult to conduct metallization with immersion times of less than 15 min. The meager nucleation on the hydrophobic polyimide surface apparently impeded metallization after CONV catalyzation. On sample catalyzed for 20 min in Fig. 6.11d, the Ni-P film was partially deposited. Autocatalytic reduction reactions in solution generally need an induction time to attain total coverage of a polyimide substrate. In our result, more than 20 min was essential as the induction time in the CONV catalyzation reaction for total Ni-P coverage on polyimide substrate.

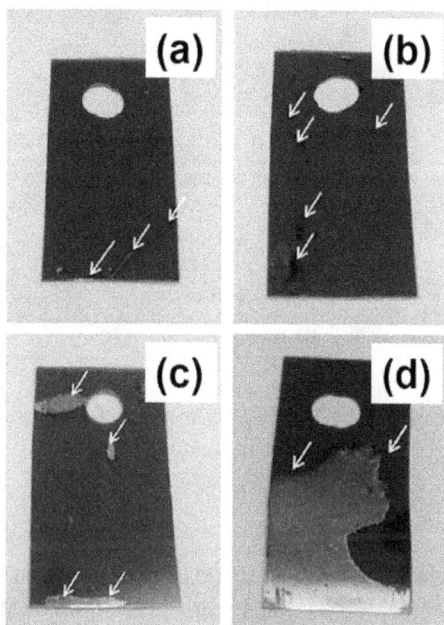

Fig. 6.11. Photographic images of metallization for 5 min on CONV
catalyzation for different times on polyimide substrate (a) 5 min, (b) 10 min,
(c) 15 min, and (d) 20 min. Metallized Ni-P parts were indicated
with white arrows.

Fig. 6.12, meanwhile, shows that catalysts are well deposited and
distributed in the film deposited within a short period of sc-CO$_2$
catalyzation. According to XRD and EDX analyses, the white particles
in the SEM images are Pd(acac)$_2$. The amount and size of these catalysts
increased and expanded a little as the catalyzation in the sc-CO$_2$
proceeded over time. The crystal growth of the catalyst commenced
when the catalyzation time reached 10 min. Compared with the films
catalyzed by the CONV method, catalyst nuclei were uniformly
deposited and dispersed. This condition may have resulted from the good
affinity of the sc-CO$_2$ with the hydrophobic polyimide surface. We also
observed the deposition of abundant catalysts on the substrate within a
relatively short time. This was principally because the diffusion
coefficient of sc-CO$_2$ media is 10 or 100 times higher than aqueous Pd
activation [8, 16, 17]. Sc-CO$_2$ diffusibility could also encourage the
Pd-complex to disperse over all of the surface-treated polyimide, thereby
facilitating uniform catalyzation growth.

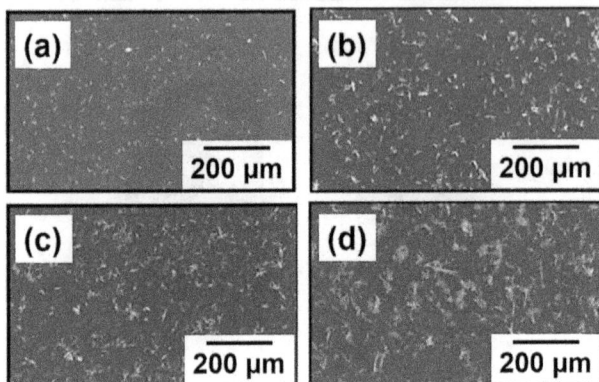

Fig. 6.12. SEM images of catalyzed polyimide surface in sc-CO_2 for different times (a) 5 min, (b) 10 min, (c) 15 min, and (d) 20 min. White particles are the catalysts deposited in sc-CO_2.

The average roughness (Ra) of the substrate washed in HCl solution for 1 min was 0.81 nm. The Ra values were 1.44 and 6.68 nm, respectively, in films metallized with CONV catalyzation for 5 min and 20 min. The Ra of the surface after CONV catalyzation increased heavily with time. In contrast, the Ra values of the surfaces after sc-CO_2 catalyzation for 5 and 20 min were 2.61 and 3.27 nm, respectively. Thus, the Ra in sc-CO_2 catalyzation increased far less than that in CONV catalyzation over time. It may have been difficult, in the films treated by CONV catalyzation, for palladium to deposit on the substrate. The hydrophobic property of the polyimide surface and the poor wettability of the polymer substrate with the $PdCl_2$/HCl solution may have led to the localization of some catalyst on the polyimide substrate at the initial stage of high density of PdCl. Thus, the palladium nuclei growing of the polymer substrate grew on the palladium nuclei initially deposited on the polyimide. $Pd(acac)_2$ nuclei deposited by the sc-CO_2 catalyzation for 5 min were well deposited on all points of the substrate. Because sc-CO_2 is a hydrophobic media, it has good wettability with the polymer substrate. The Ra in the samples catalyzed for 20 min in sc-CO_2 were not increased as much as those treated by CONV catalyzation. Thus, we suggest that the nucleation density on the polyimide surface was higher in sc-CO_2 catalyzation than in CONV catalyzation. To precisely clarify catalyst nuclei growth in different catalyst reaction times, we quantitatively analyzed the catalyzation nucleation in sc-CO_2 using an SEM image.

Fig. 6.13 shows the areas over which catalyst particles occupy a polyimide substrate catalyzed in sc-CO_2 at various catalyzation times.

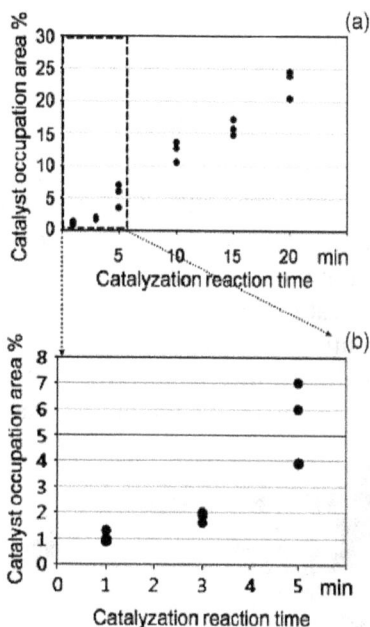

Fig. 6.13. Catalyst occupation area on polyimide substrate catalyzed in sc-CO_2
with various catalyzation times (a) and a dotted box in (a)
was expanded on (b).

Catalyst particles on polyimide were analyzed with visual point counting after ethanol cleaning. This occupation area shows amount of catalyst penetrating remained in polyimide. As the figure shows, the substrate area occupied by the catalyst increased as the reaction time increased. Here we consider the nucleation on polyimide in an "open system," where new nuclei continue to nucleate on the surface during growth of the nucleated catalysts [18]. At catalyzation times from 1 to 3 min, the area occupied by the catalyst ranged from 1 to 3 %. At a catalyzation time of 5 min, the area occupied by the catalyst increased to about 7 %. After 5 min of catalyzation, the area occupied by the catalyst on polyimide substrate increased quasi-linearly. The area of the substrate occupied by the catalyst rapidly increased from 3 min to 5 min. Lau discussed the nuclei density and the nucleus size in an autocatalytic

181

reaction of Pd on TiN substrate [18]. The autocatalytic reaction of Pd had three stages, namely, growth, secondary nucleation, and ripening, all of which strongly influenced the surface roughness. There was not observed such a behavior on the case of this catalyst reaction in sc-CO_2. It may be that the catalyzation reaction in sc-CO_2 is not an autocatalytic reaction, but a deposition reaction of the catalyst in the amphiphilic media to the polymeric substrate and the catalyst, without an electrochemical reduction. These results affect Ni-P coverage ratio.

Fig. 6.14 shows OM images of Ni-P thin film surface after sc-CO_2 catalyzation time from 1 to 5 min. The Ni-P thin film after catalyzation for 1-3 min formed an island structure (Figs. 6.14a and b) and then merged to form a continuous structure after catalyzation for 5 min (Fig. 6.14c). This indicates that the polyimide surfaces catalyzed after 1-3 min were not continuously deposited with Ni-P thin film and the uncoated parts formed porosities of the Ni-P film. In catalyzation for 5 min, complete coverage by the Ni-P thin film was obtained.

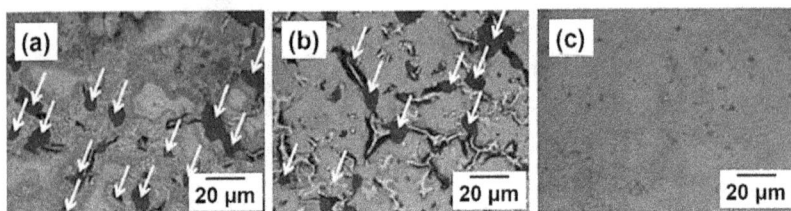

Fig. 6.14. OM image of Ni-P thin film surface with different sc-CO_2 catalyzation times (a) 1 min, (b) 3 min, and (c) 5 min. Porosities on Ni-P film were indicated with white arrows.

Fig. 6.15 also shows Ni-P coverage ratios at different catalyzation times from 1 min to 5 min. Complete coverage was not attained with catalyzation from 1 to 3 min, whereas it was obtained with catalyzation for 5 min in sc-CO_2. Ni-P coverage ratio was drastically increased from 21 to 91 % in catalyst reaction timefrom1 to 3 min, and reached full coverage over 5 min. Here we observe a dramatic change from 3 to 5 min, compared to the result shown in Fig. 6.13. This period is considered to be a critical point for full coverage. This result indicates that an area covered by critical catalyst is a necessary precondition for continuous coverage of the Ni-P thin film. Moreover, the catalyst sizes are also thought to influence the coverage. Though the areas occupied by the catalyzation changed only moderately with the change in the catalyzation time from 1 to 3 min, the coverage of Ni-P catalyzed for 3 min was about 3 times higher than that catalyzed for 1 min. On this basis, we know that the increase of the catalyst size with the catalyzation time influences the metallization coverage. The catalyst size and the area of the polymer substrate occupied by the catalyst are indispensable to the completely continuous coverage of the Ni-P film over the polymer substrate.

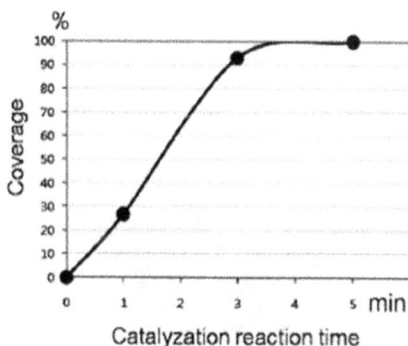

Fig. 6.15. Ni-P coverage on different sc-CO_2 catalyzation times.

6.2.1.3. Sc-CO_2 Catalyzation and Sc-CO_2 Metallization

The substrate was washed with acetone and rinsed in deionized water before each reaction. Grease was removed from the sample by successive dipping in a 10 wt.% solution of NaOH and a 10 wt.% solution of HCl, followed by rinsing in deionized water. The maximum working temperature and the maximum pressure were 150 °C and 50 MPa, respectively. Similar metallization reaction system was used in this section. An OM and FE-SEM were used to observe the surfaces of the metallized Ni-P films. The Atomic Force Microscopy (AFM) survey of catalyzed polyimide surfaces was performed using SPA-400, Seiko Instruments, Inc. at ambient conditions. Five different positions on each sample were scanned to ensure that the images obtained were representative of the surface structure. Only one location was then examined in detail. Especially we excepted scratch area for comparing aqueous and sc-CO_2 media effect. Two types of catalyzation procedure were performed. Identical CONV catalyzation and sc-CO_2 catalyzation were conducted. The reaction temperature and pressure were 80 °C and 15 MPa, respectively. Catalyzed polymer substrates were metallized in Ni-P metallization solution by two types of metallization reaction. In the CONV metallization, catalyzed polyimide substrate was immersed in a Ni-P solution and then a Ni-P thin film was fabricated. The metallization in the sc-CO_2-assisted Ni-P metallization was conducted at 80 °C and 15 MPa. A small amount (300 µl) of polyoxyethylene lauryl ether ($C_{12}H_{25}(OCH_2CH_2)_{15}OH$) was used as a non-ionic surfactant for the mixing of the metallization solution with sc-CO_2. The volume ratio of the CO_2 in the autoclave was 0.6.

Fig. 6.16 shows AFM images of a Kapton surface catalyzed by CONV catalyzation in a $PdCl_2/SnCl_2$ mixture solution at room temperature, and by catalyzation in sc-CO_2 at a reaction temperature and pressure of 80 °C and 15 MPa, respectively.

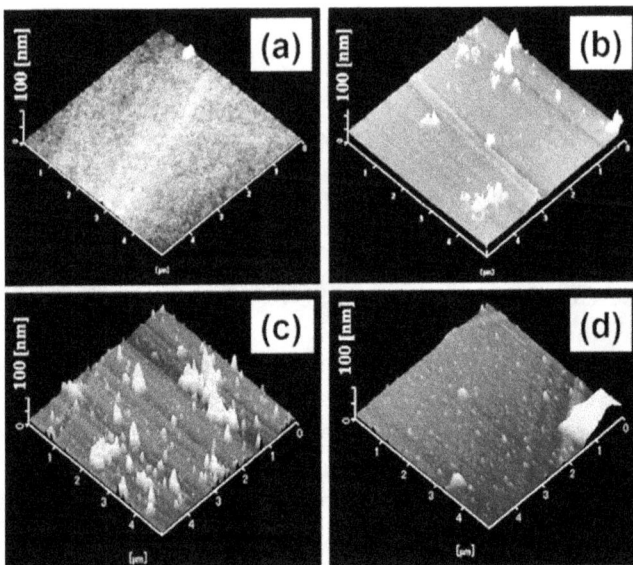

Fig. 6.16. AFM images of catalyzed Kapton surface by (a) CONV catalyzation for 5 min, (b) catalyzation in sc-CO_2 for 5 min, (c) CONV catalyzation for 20 min, (d) catalyzation in sc-CO_2 for 20 min.

The reaction times in both methods were 5 min and 20 min. In the CONV catalyzation shown in Fig. 6.16a, a lone palladium nucleus can be observed as a white point on the polyimide surface at 25 µm². In the sc-CO_2 catalyzation shown in Fig. 6.16b, as many as 20 palladium nuclei are observed in the same area. Fig. 6.16c is by CONV catalyzation for 20 min. This figure shows that there are lots of palladium nucleation which were aggregated. In the sc-CO_2 catalyzation shown in Fig. 6.16d, palladium nuclei were uniformly deposited and distributed on polyimide substrate. In the CONV catalyzation for 5 min, the Pd nuclei in the sc-CO_2-catalyzed polyimide was triple the height of the lone nuclei observed in the CONV catalyzation. The catalyzation solution was incompatible with the hydrophobic surface of the polyimide substrate. Hence the nucleation of Pd on the polyimide substrate was difficult in the aqueous solution. The compatibility between the polyimide and sc-CO_2 was otherwise good. The Pd nuclei were easily deposited on the polyimide in sc-CO_2. The abundance of Pd nuclei facilitated the deposition of the Ni-P thin film onto the polyimide surface. Hence, the sc-CO_2-catalyzed polyimide surface was easily metallized by the Ni-P film in the metallization reaction. In case of CONV catalyzation for 20 min, there are also lots of catalyzation on polyimide substrate. However, many white points are aggregated. This indicates that palladium grows on palladium nucleus which was firstly deposited on polyimide. Thus some large nuclei were observed.

On the other hand, the nuclei in sc-CO_2 catalyzation were uniform and well scattered on polyimide substrate. This difference is suggested to come from the compatibility of sc-CO_2 with hydrophobic polyimide. Sc-CO_2 diffusivity could also make Pd-complex move to all the part of surface treatment of polyimide, and would facilitate uniform catalyzation growth. Surface roughness data confirm this suggestion. Firstly Ra of the substrate washed in HCl solution for 1 min was 0.81 nm. In case of CONV catalyzation for 5 min and 20 min, Ra values were 1.44 and 6.68 nm, respectively. Ra of the surface after CONV catalyzation heavily increased with time. However Ra values of that after sc-CO_2 catalyzation for 5 and 20 min were 2.61 and 3.27 nm, respectively. Thus, as compared with the result of CONV catalyzation, Ra in sc-CO_2 catalyzation did not increase so much with time. In case of CONV catalyzation, it could be difficult for palladium to deposit on substrate and some catalyst firstly could be localized on polyimide substrate, because polyimide surface has hydrophobic property and wettability of polymer substrate with $PdCl_2/SnCl_2$ mixture solution is bad. Thus palladium nuclei of polymer substrate grew on palladium nucleus which was firstly deposited on polyimide. Pd nuclei by sc-CO_2 catalyzation for 5 min were well deposited on all point of substrate. Since sc-CO_2 is hydrophobic media,

185

it has good wettability with polymer substrate. For 20 min catalyzation in sc-CO_2, Ra was not increased so much as compared with CONV catalyzation. Thus we suggest that the nucleation density on polyimide surface in sc-CO_2 catalyzation was higher than in CONV catalyzation. More evidence will be needed to formulate a detailed mechanism for our process. Nevertheless, our proposed sc-CO_2 catalyzation produced even more outstanding results than CONV catalyzation.

In the first experiments, CONV metallization was carried out after CONV catalyzation. Fig. 6.17a shows small, island-like formations metallized with Ni-P. The meager nucleation on the hydrophobic polyimide surface apparently impeded CONV metallization after CONV catalyzation. But when the sc-CO_2-catalyzed polyimide substrate was immersed in the Ni-P metallization solution (Fig. 6.17b), the entire substrate surface was metallized with Ni-P thin film. On the basis of this result, we concluded that sc-CO_2 catalyzation works well with hydrophobic polymer substrate.

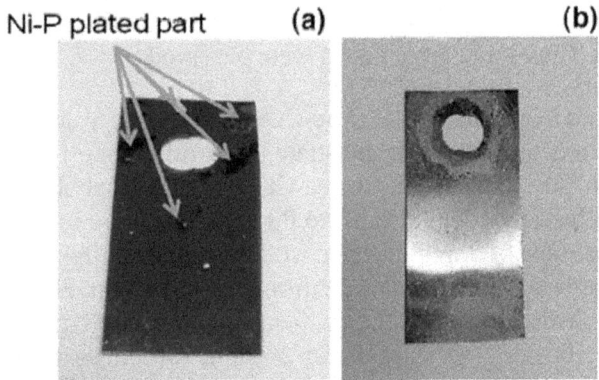

Fig. 6.17. Photographic images of CONV metallization on CONV catalyzation (a) and catalyzation in sc-CO_2 (b).

Fig. 6.18 shows an OM image and SEM image of a Ni-P thin film surface after catalyzation of CONV metallization. Figs. 6.18a and b show numerous peeled sections and nodules on the Ni-P thin film. The peeling resulted from the hydrogen bubbles generated by the side reaction of the metallization on the polyimide substrate. In the CONV metallization, the hydrogen gas generated by the side reaction led to the formation of pinholes and voids on the Ni-P thin film. Judging from this result, we concluded that sc-CO_2 catalyzation facilitates Pd nucleation on hydrophobic polyimide surfaces, but never hinders the generation of voids and cracks.

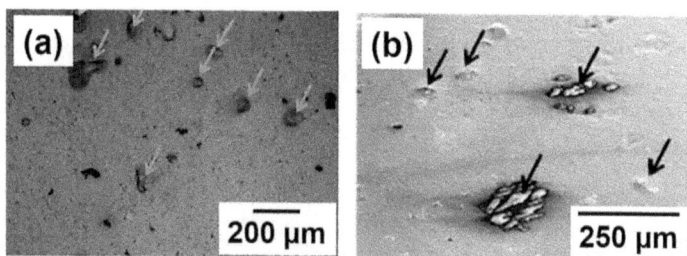

Fig. 6.18. OM image (a) and SEM image (b) of Ni-P thin film surface after catalyzation of CONV.

Moreover, sc-CO_2-assisted Ni-P metallization was conducted after the sc-CO_2 catalyzation. Fig. 6.19 shows an OM image and SEM image of a Ni-P thin film surface formed by sc-CO_2-assisted Ni-P metallization after sc-CO_2 catalyzation.

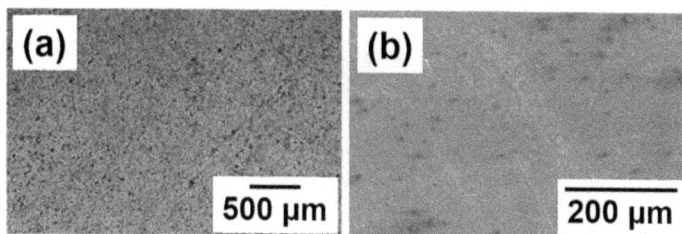

Fig. 6.19. OM image (a) and SEM image (b) of sc-CO_2-assisted Ni-P metallization after catalyzation in sc-CO_2.

Here, in order to discuss the effect of metallization using sc-CO_2 on voids and nodules of the metallized film, the metallized films with the thickness of about 0.5 μm were fabricated. Fig. 6.19a and b show that the process resulted in a uniform thin film without peeled sections on the view point of macroscopic property and porosity. In Fig. 6.18, there were voids and nodules, and observed small convex parts about 1 μm, which were originated from the rough surface of Kapton. Also there observed only the convex parts in Fig. 6.19. Sc-CO_2 dissolves hydrogen gas and thus decreases the peeling in sc-CO_2-assisted metallization. Sc-CO_2 has high diffusivity and high miscibility with H_2. This leads to the dissolution of the generated H_2 in sc-CO_2 and thus decreases the peeling on the polyimide substrate [19].

Fig. 6.20 shows cross sectional SEM images and EDX images of the interface between the Ni-P thin film and polymer being catalyzed in sc-CO_2.

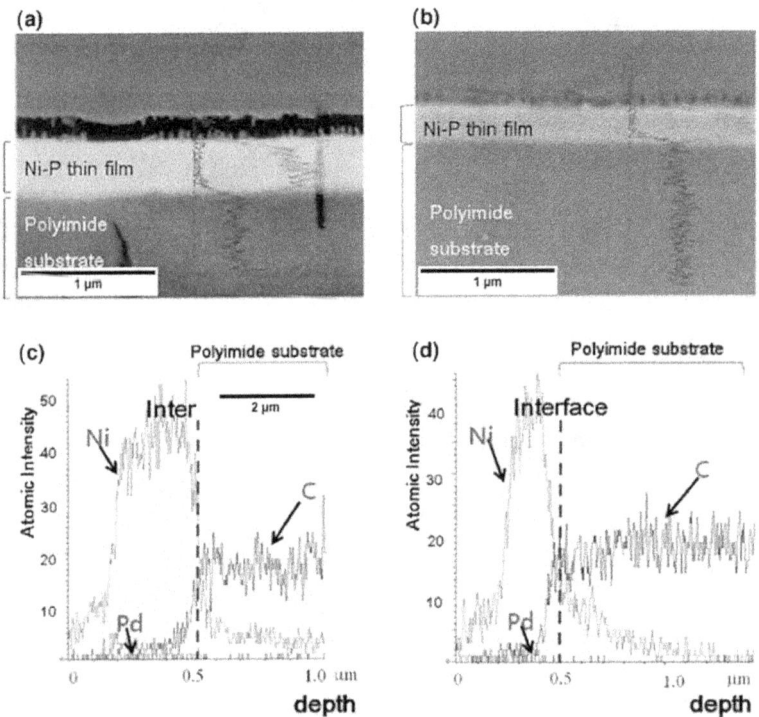

Fig. 6.20. Cross sectional SEM images (a) and (b) and EDX analysis (c) and (d) of interface between Ni-P and polymer substrate which were catalyzed by catalyzation in sc-CO_2. (a), (c) by CONV metallization; (b), (d) by sc-CO_2-assisted Ni-P metallization.

The intensity of Ni can be observed at a depth of 200 nm inside the polyimide in sc-CO_2-assisted Ni-P metallization, whereas no Ni atom is observed inside the polyimide in CONV metallization. This indicates that Pd nucleates at a depth of 200 nm in sc-CO_2 due to the high transport property of the sc-CO_2. The compatibility between the hydrophobic substrate and the reaction media is also very important for the immersion of the Ni-P metallization solution inside the polyimide. In CONV metallization, the substrate was metallized after the palladium nuclei were deposited onto the polyimide surface. Packed air or diffusivity, however, prevented the Ni-P solution from reaching deeply into the recessed parts. In sc-CO_2-assisted Ni-P metallization, the Ni-P reached into, and reacted within, the deeply recessed parts of the substrate. Interfaces such as interlocking structures were thus formed via the sc-CO_2 catalyzation [20].

6.2.1.4. Effects of Sc-CO_2 Catalyzation in Sc-CO_2 Metallization

Similar catalyzation and metallization methods were carried out in this section. Next, the same catalyzation procedure was performed in 40 ml of n-hexane solvent (Sigma Aldrich, USA) with 0.25 g Pd(acac)$_2$ at 80 °C for 30 min. An OM and FE-SEM were used to observe the surfaces of the metallized Ni-P films. Samples were coated by 5 nm osmium coater and embedded in epoxy resin for SEM-EDX observation. The cross section of the Ni-P sample was observed by SEM. EDX was used to observe the Ni-P metal penetration within the polyimide substrate. Point analysis was performed on a cross sectional plane in the depth direction from the surface. Cross sections of Ni-P thin film on polyimide were prepared using an ultramicrotome (Leica EM UC6) and examined via SEM and EDX.

The mathematical theory of diffusion through isotropic substances was first studied by Fick [21]. This theory is based on the hypothesis that the rate transfer of a diffusing substance through unit area of a section is proportional to the concentration gradient measured normal to the section, i.e.,

$$J = -D\frac{\partial C}{\partial x},$$
(6.1)

where D is the diffusion coefficient, C is the concentration of the diffusion substance, x is the coordinate perpendicular to the section and J is the flux per unit of area. This is known as Fick's first law of diffusion. The differential equation of diffusion, known as Fick's second law derives from the first one and if the diffusion is one dimensional, i.e., if there is a gradient concentration in only one direction (along the x-axis), the amount of diffusion substance in the element is given by the expression;

$$\frac{\partial c}{\partial t} = D \frac{\partial^2 c}{\partial x^2}. \tag{6.2}$$

A particular solution of the differential equation (Fick's second law of diffusion) describing the time dependence of the diffusion material out of the sample, (in our case, the equation is given in a semi-infinite system which is usually applied to one dimensional diffusion of gas elements in a material): the material is in contact with sc-CO_2, and the solute concentration at the surface is kept constant (C_0). The boundary conditions are given in the forms, C = 0, for x > 0, at t = 0, C = C_0, for x = 0, at t > 0, where 'C_0' is a sc-CO_2- element concentration at the surface and is constant regardless of time, 'C' is a sc-CO_2 element concentration at a distance 'x' from the surface and 't' is time. The concentration at t = 0 is assumed to be 0. 'C' can be obtained using the error function (erf) which is expressed in the form,

$$\frac{C}{C_0} = 1 - erf\left(\frac{x}{2\sqrt{Dt}}\right). \tag{6.3}$$

In this study, the relationship between the penetration of Ni-P ions and reaction time for catalyzation in sc-CO_2 is discussed with reference to the above Eq. (6.3). The diffusion coefficient of CO_2 into polyimide in both temperature and pressure beyond critical point is necessary for the calculation of CO_2 diffusion into polymer, but there is no report about diffusion study of CO_2 into the polyimide in high pressure region. Sato et al. have studied the diffusion behavior of CO_2 into various polymers as functions of temperature and pressure and especially in supercritical state [17, 22]. They reported that the diffusion coefficient of CO_2 into polymers increases with increases of pressure and temperature and moreover the diffusion coefficient of sc-CO_2 into polymer such as poly(butylene succinate) (PBS) and poly (butylene succinate-co-adipate) (PBSA) are about 1×10^{-9} (m^2/s) at 10-12 MPa and 120 °C - 180 °C. Strictly, the diffusions of CO_2 in PBS and PBSA are different from that into polyimide. But we believe that the magnitude of the parameters about diffusion into polymer can be the same and so we used the parameter of diffusion of PBS in the studies by Sato et al. because the parameters are quantitative in the region of 120 °C and 12.3 MPa. Fig. 6.21a shows calculated CO_2 fraction in polymer by Eq. (6.3) using diffusion coefficient of CO_2 in PBS about 1.23×10^{-9} (m^2/s) at 120 °C,

12 MPa with various sc-CO_2 catalyzation reaction times. Horizontal line is penetration depth into polymer, and vertical line is ratio of carbon dioxide concentration. The concentration ratio of penetrated CO_2 is dependent on reaction time. It is not general that there is linear relationship between media concentration and penetration depth for Eq. (6.3), as shown in Fig. 6.21. In this case, we suggested that sc-CO_2 has high diffusion coefficient in polymer and thus there is such linear relationship between CO_2 concentration and penetration in polyimide in limited region of small penetration depth.

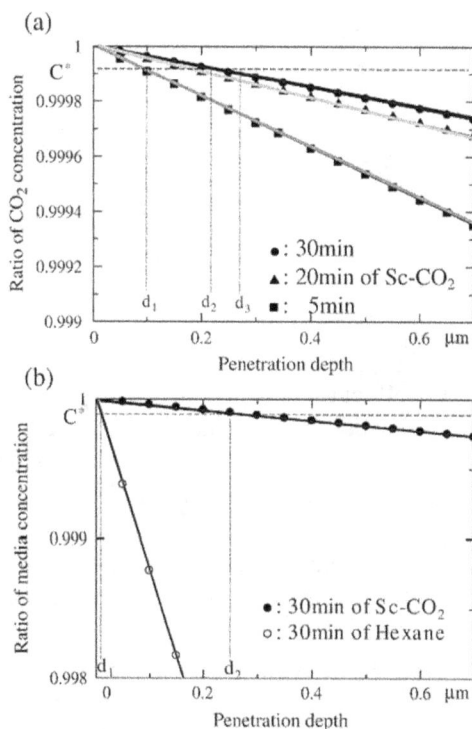

Fig. 6.21. Theoretical results with various reaction times (a), calculation results of sc-CO_2 and hexane media (b).

There can be a linear relationship between the ratio of CO_2 and Pd concentrations because Pd organo-complex dissolved in CO_2. Thus penetration depth of Pd organo-complex is determined with sc-CO_2 diffusion. Moreover, in consideration of metallization reaction, we proposed C* which is the ratio of CO_2 concentration into polymer to transport critical Pd concentration enough to react Ni-P metallization and to deposit Ni-P. By using C*, penetration depths d_1, d_2, and d_3 were determined at CO_2 fraction of 0.9999. Relation of penetration depth and root square shows linear line in Fig. 6.22. This calculated result will be discussed with experimental result.

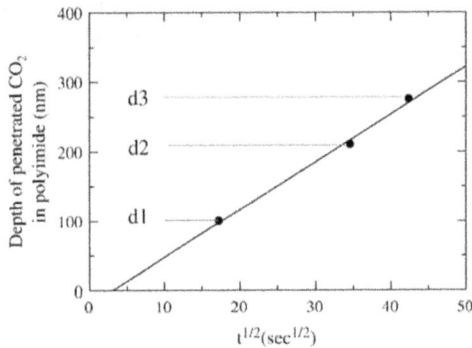

Fig. 6.22. Theoretical results with various reaction times and depth of penetrated CO_2 in polyimide.

In order to clarify the effect of diffusion coefficient on the CO_2 diffusion into polyimide, we selected hexane which solvent properties are known to be similar to those of sc-CO_2. However, there is no report about diffusion study of hexane into the polyimide. Jiang et al. studied diffusion behavior of n-hexane in polyisobutylene (PIB) and we use these parameters for the calculation of diffusion into polymer to qualitatively discuss the diffusion behavior of hexane into polyimide. Fig. 6.21b shows calculated media fraction of sc-CO_2 and hexane in polyimide using diffusion coefficient of n-hexane in PIB about 1.18×10^{-12} (m^2/s) at 90 °C [17, 22]. Fig. 6.21b shows calculation results with 30 min reaction time of sc-CO_2 and hexane respectively, diffusion depth result shows huge difference, due to the diffusion coefficient. These calculated results indicate that Pd complex in hexane can hardly penetrate into polyimide and will be discussed with experimental result.

Fig. 6.23 shows the Ni-P thin film on polyimide using sc-CO_2-assisted metallization for 30 min after catalyzation in sc-CO_2 for 5, 20, and 30 min. We performed a set of experiments under fixed conditions for the sc-CO_2-assisted metallization in order to study the effects of the catalyzation reaction time in sc-CO_2 on the conformal and interface stability of the Ni-P thin film. The polyimide was completely metallized with the Ni-P metal film in every sample. Compared with the Ni-P surfaces metallized by the CONV metallization method, the surfaces fabricated using sc-CO_2 were uniformly metallized and free of pinholes, nodules, and blisters in both the macro and micro-sized observations. The cross sectional morphology, however, differed under SEM observation.

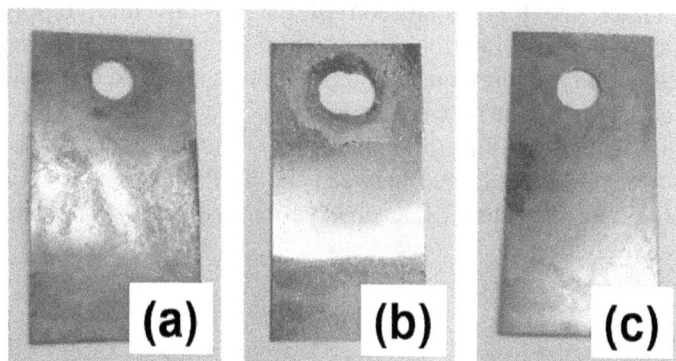

Fig. 6.23. Photographic images of Ni-P thin film on polyimide using sc-CO_2-assisted metallization for 30 min after various catalyzation times in sc-CO_2 (a) 5, (b) 20 and (c) 30 min.

Fig. 6.24 shows cross sectional SEM images on the interface between the Ni-P thin film and polymer using sc-CO_2-assisted metallization for 30 min after catalyzation in sc-CO_2 for 5, 20, and 30 min. Figs. 6.24a and b reveal several small cracks and small round particles on the cross sectional plane of the Ni-P thin film. In contrast, Fig. 6.24c shows good filling on the cross sectional plane. The increased catalyzation time apparently resulted in a far more uniform interface structure. These results demonstrate that the nucleating sites for Ni-P crystal growth increased as the catalyzation in sc-CO_2 time increased. In contrast to the uniform film shown in Fig. 6.24c, the photographs in Figs. 6.24a and b reveal many defects resulting from the low catalyst density of the nucleating sites, in the polyimide substrate. Metallization is initiated on the catalyst particles randomly distributed on the substrate, and the initial metal structure is largely determined by the surface morphology and surface roughness of the film [18]. Patterson et al. reported that

193

metallized Cu morphology and grain structure were dependent on the Pd activation process and the control of the subsequent Cu plating [23, 24]. Nakahara, meanwhile, found that the properties of the seed layer determined the electroless Cu morphology [25]. The Ni-P micro size morphology is important for mechanical characterization, as small voids or cracks in micro-sized areas affect the reliability of materials. Next we will discuss the penetration of the Ni-P, i.e., the anchoring property conferred by the Ni-P growth into the polyimide in the depth direction. Some results and reviews have clarified sc-CO_2 impregnation properties on polymer substrates with weight variation [26, 27]. Here, however, we must characterize the penetration of the Ni-ions into a polyimide substrate, as the Ni-P interlocking structure plays a key role in securing interface stability. In previous sections, the amounts of Pd that penetrated into the polymer matrix were almost too scanty to be detected for analysis. To solve this dilemma, we conducted cross sectional observations of the interface between the Ni-P and polymer substrate with EDX. We selected EDX observation with point analysis from Ni-ions in polyimide.

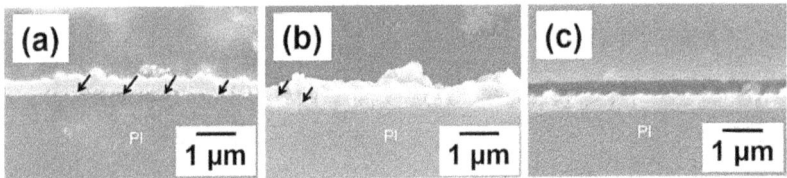

Fig. 6.24. Cross sectional SEM images of interface between Ni-P and polymer using sc-CO_2-assisted metallization various catalyzation times in sc-CO_2 for (a) 5, (b) 20 and (c) 30 min.

Fig. 6.25 shows EDX analysis images of Ni-ions on cross sectional planes between the Ni-P thin film and polyimide substrate after catalyzation for 5, 20, and 30 min in sc-CO_2. We analyzed penetrated Ni-ions in polyimide with EDX from the interface between the Ni-P thin film and polyimide substrate. As shown in Fig. 6.22, there is liner relationship between CO_2 fraction and penetration depth. The concentration of deposited Pd complex can be proportional to CO_2 fraction in polyimide and thus the deposited Ni is also directly influenced by deposited Pd complex [28]. Hence observed Ni-ions in polyimide can have linear relationship with penetration depth and so linear approximation line was plotted in Fig. 6.25. In previous work [29], the intensity of Ni can be observed at a depth of 200 nm inside the polyimide in our novel electroless plating method. No Ni atom is observed in the polyimide by CONV metallization. But, in CONV metallization after catalyzation in sc-CO_2, the intensity of fewer than 10 % Ni was shown at a depth of 500 nm inside the polyimide in sc-CO_2-assisted metallization. Thus we consider that Ni peaks of under 10 % were within the error range. Fig. 6.25a shows that penetrated Ni peaks were penetrating to about 80-90 nm. Fig. 6.25b shows penetrated Ni peaks to about 190-210 nm. Fig. 6.25c shows strong Ni peaks penetrating deeply to 290-300 nm.

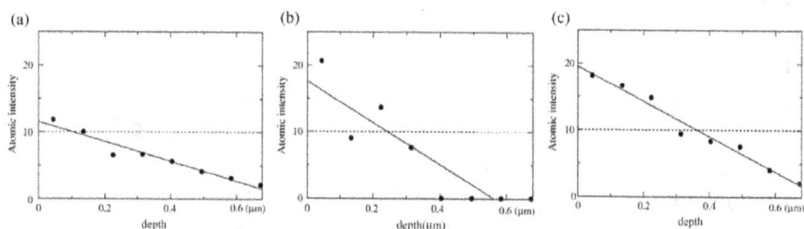

Fig. 6.25. EDX analysis of Ni-ions on the cross sectional plane between the Ni-P film and polyimide which were catalyzed in sc-CO_2 for various reaction times (a) 5, (b) 20, (c) 30 min by sc-CO_2-assisted metallization.

Fig. 6.26 plots the penetration of the Ni-ions in polyimide vs. the square root of various catalyzation times in sc-CO_2 in films fabricated using sc-CO_2-assisted metallization for 30 min. The depth of the penetrated Ni-ions and the square root of catalyzation times in sc-CO_2 were confirmed to be linearly related, and thus to be in conformance with the calculated data of Fig. 6.22 [30], although the calculated results in Fig. 6.22 were based on the diffusion coefficient in PBS. Fick's second law of diffusion describes the time dependence of the diffusion of the material into the sample and we considered that the magnitude of diffusion coefficient of CO_2 into PBS might be close to that of polyimide. In this result, it is interesting that there exists a pseudo incubation time for about 36 s. We consider that this pseudo incubation time means the time of one nucleation concentration of Pd catalyst for Ni-P metallization into polyimide. After the pseudo incubation period, impregnation of Ni with 10 % peak intensity was progressed by reaction time. We suggest that Pd organo-complex nucleation can be homogeneous. Bulk structure of polyimide is constructed with plenty of polymer chains and thus there are no special points for nucleation. Much vacancy exists between polymer chains for penetrating of CO_2. Fig. 6.26 shows the experimental results to be in good agreement with the theoretical results obtained d_1, d_2, d_3 in Fig. 6.22. Hence the impregnation reaction of Ni by our novel method is CO_2 diffusion controlling reaction. Moreover, these deeply penetrated Ni-ions are a key to achieving the good stability between the Ni-P metal film and polyimide. These results mean that our proposed C* could exist in the impregnation reaction of Ni-P into polyimide by our proposed electroless plating method. We should discuss the effect of reaction condition in the electroless plating reaction using sc-CO_2. We selected 80 °C for the catalyzation in CO_2 solution of Pd catalysts and the sc-CO_2-assisted metallization because 80 °C is good enough for both processes. In the catalyzation in CO_2 solution, the impregnation depth of Pd catalysts into polymer increases with the increase of temperature. On the other hand, the process of the Ni-P impregnation could change from reaction controlled to diffusion controlled with temperature increasing. In the case of this study, the temperature of 80 °C is high enough for the diffusion controlled impregnation. Thus, lower temperature condition might cause the impregnation to be reaction controlled. Ohshima et al. reported an electroless plating process on thermoplastic polyamide, which consists of sc-CO_2 impregnation, that a Pd catalyst is infused into polymer by solvency and plasticization powers of sc-CO_2, and metallization reaction in the presence of ethanol and CO_2 [6]. The addition of ethanol and CO_2 to the solution enhanced the diffusivity of the solution due to solvent swelling of the polymer. They also concluded that in the reaction process, the reaction rate is controlled by the diffusion of metal ions in the ternary mixture of metallization solution, ethanol and CO_2. In the case of high diffusivity, the metal ion penetrates into the polymer from the surface before the metal film completely coats the surface. We agree their conclusion that the diffusivity of metal ions in the metallization solution with sc-CO_2 controls the metallization reaction on polymer, although the metallization reaction media and the polymer substrate are different from those in our reaction system. Our study also clarified that the diffusivity of sc-CO_2 carrying Pd catalyst controlled the depth that the metallization reaction occurred into polymer. Both studies can give the understanding of the impregnation of the metallization into polymer by using dense CO_2.

To ascertain the effects of the sc-CO_2 diffusion coefficient on high transport properties, we fabricated a Ni-P thin film using sc-CO_2-assisted metallization after catalyzation in hexane media. The photographs in Fig. 6.27 show Ni-P thin films fabricated by sc-CO_2-assisted metallization after catalyzation in hexane (a) and catalyzation in sc-CO_2 (b) for 30 min. With catalyzation in hexane, the Ni-P thin film was fabricated. With catalyzation in sc-CO_2, the film was completely metallized. In our previous result, the chemical affinity made it difficult to deposit catalyst on a hydrophobic polymer substrate by CONV catalyzation with a mixed $PdCl_2/SnCl_2$ solution. In our present experiment, hexane had good chemical affinity with the polyimide substrate. Palladium catalyst was easily deposited on the polyimide substrate in both media. Yet under microscopic observation, the Ni-P structure fabricated with catalyzation in hexane was marred with many surface defects.

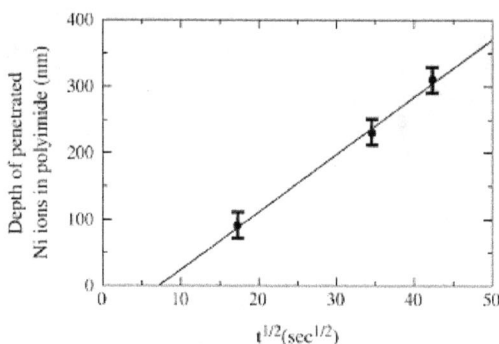

Fig. 6.26. Plot of depth of the penetrated Ni-ions in polyimide vs. the square root of various catalyzation times in sc-CO_2-assisted metallization for 30 min.

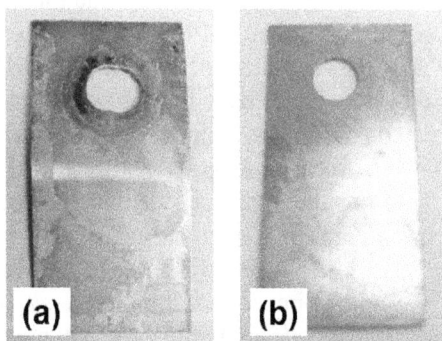

Fig. 6.27. Photographic images of Ni-P thin film by sc-CO_2-assisted metallization after catalyzation in hexane (a) and catalyzation in sc-CO_2 (b).

Fig. 6.28 shows OM and SEM images of Ni-P thin film surfaces using sc-CO_2-assisted metallization after catalyzation in hexane (a), (c) and catalyzation in sc-CO_2 (b), (d).

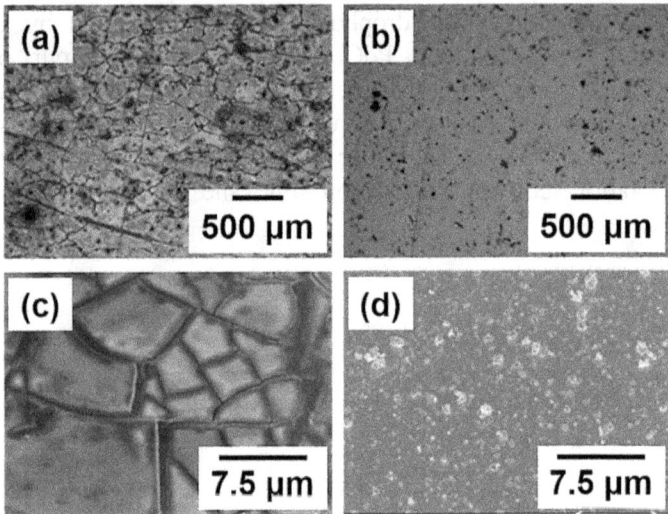

Fig. 6.28. OM and SEM images of Ni-P thin film using sc-CO_2-assisted metallization after catalyzation in hexane (a), (c) and catalyzation time in sc-CO_2 (b), (d).

Figs. 6.28a and c reveal numerous cracks on the Ni-P thin film. These cracks were likely caused by the difference in the volumetric thermal expansion coefficients between Ni-P thin film and the substrate polymer. The thermal expansion coefficients of Ni-P and polyimide are 1.3×10^{-5} cm/cm K and 2.0×10^{-5} cm/cm K, respectively. This difference caused a large internal force between the metal and the polymer. It formed cracks in the metal film when cooling the samples down from the metallization temperature of 80 °C to room temperature after metallization. In Figs. 6.28b and d, we find that the process resulted in a uniform thin film without any apparent peeling or cracking in the macroscopic and microscopic views. The Ni-P growth from the existence of Pd nucleates into polyimide is thought to improve the interface stability on the Ni-P film surface via an anchoring effect. In the case of catalyzation in hexane, palladium catalysts were easily deposited on the polyimide. But the polyimide, a flexible material, was swollen with the sc-CO_2-assisted metallization. Moreover, the polyimide was easily bent by the force of the stirring in the sc-CO_2-assisted metallization solution in the reaction cell. In the specimens without Ni-P growth in the depth direction, cracks easily propagated during the stages of Ni-P growth. Sc-CO_2 has a higher diffusivity than hexane. This high diffusivity leads to the penetration of the Pd catalyst and Ni-P metallization solution into the polyimide substrate, that is, Ni-P growth in the depth direction. Thus, the interface stability between the Ni-P and polyimide is improved.

Fig. 6.29 shows cross sectional SEM images of the interface between Ni-P and polyimide of the metallized sample using sc-CO_2-assisted metallization after catalyzation in hexane (a) and sc-CO_2 (b). Figs. 6.29a and b both show stable cross sectional planes. The uniform interface of the cross sectional plane structure after catalyzation in hexane tells us that the catalyst density was sufficient for fine Ni-P growth. But under the macroscopic observation, we find an abundance of cracks and peeled sections. In case of catalyzation in hexane (Figs. 6.28a and c), the incubation time is sufficient for Ni-P full coverage, but the comparative lack of Ni-P growth in the depth direction leads to a weaker anchoring structure for interface stability. To characterize the penetrated Ni-ions in the polyimide, we also observed them in cross sectional planes.

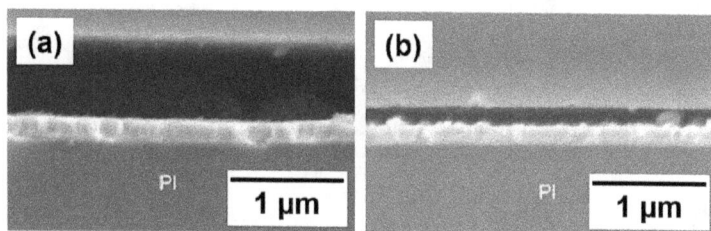

Fig. 6.29. Cross sectional SEM images of interface between Ni-P and polymer using sc-CO_2-assisted metallization after catalyzation in hexane (a) and sc-CO_2 (b).

Fig. 6.30 shows profile of Ni contents along the depth direction, analyzed by EDX on cross sectional plane between the Ni-P and polyimide using sc-CO_2-assisted metallization after catalyzation in hexane (a) and sc-CO_2 (b). The results here show the effects of the diffusion coefficient. Hexane and sc-CO_2 share similar electrochemical properties, yet their diffusion coefficients greatly differ. The catalyzation in hexane, the intensity of penetrated Ni-ions is very low (Fig. 6.30a). On the other hand, the sc-CO$_2$ induces Pd nucleation at a depth of 300 nm during the process of catalyzation in sc-CO$_2$. These experimental results of Fig. 6.30 are in good agreement with the theoretical results of obtained d$_1$ (10 nm) and d$_2$ (275 nm) for 0.9999 ratio of CO$_2$ concentration in Fig. 6.21b. Thus this experiment also supports that the impregnation reaction of Ni by our novel method is CO$_2$ diffusion controlling reaction. The differences in diffusion coefficients lead to differences in film properties, such as the differences in cracking shown in Fig. 6.28. These data indicate the following features of the Ni-P thin film fabricated with hexane media: the Pd catalyst was deposited only on the polyimide surface, not inside the polymer, and the Ni-P metals grew only on the polyimide surface, not from inside the polymer. The penetrated Ni-P particles confer good interface stability by providing an anchoring effect between the Ni-P thin film and polyimide substrate. According to these results, an adequate catalyst density and adequate level of Ni-P metal growth in the depth direction of the surface of the polymer substrate are crucial for the suppression of cracks.

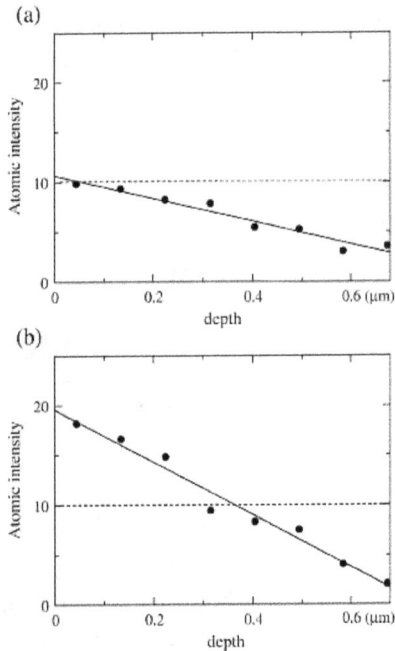

Fig. 6.30. EDX analysis images of Ni-ions on the cross sectional plane between the Ni-P and polyimide using sc-CO$_2$-assisted metallization after catalyzation in hexane (a) and sc-CO$_2$ (b).

6.2.2. Nickel-Phosphorus Metallization on Nylon 6,6

Similar high pressure apparatus and reaction chamber was used in this section. Nylon 6,6 textile was hooked from a plastic-coated platinum wire in the reaction chamber to conduct the sc-CO_2 catalyzation and the sc-CO_2-assisted metallization. No pretreatment of the Nylon 6,6 textile was conducted in this study. Two types of the catalyzation were performed. The catalyzation was performed either by the CONV catalyzation or the sc-CO_2 catalyzation. The sc-CO_2 catalyzation was performed with 2 g/l of Pd(acac)$_2$. The temperature for both the CONV catalyzation and the sc-CO_2 catalyzation was 80 °C, and the pressure for the sc-CO_2catalyzation was 15 MPa. The catalyzation for the CONV catalyzation was 10 s. Three values of catalyzation time for the sc-CO_2 catalyzation were used, which were 20, 60, and 120 min. After the catalyzation process, the metallization was conducted, which was either the CONV metallization or sc-CO_2-assisted metallization. For the CONV metallization, the catalyzed Nylon 6,6 textile was immersed in Ni-P solution. For the sc-CO_2-assisted metallization, the catalyzed textile was immersed in the sc-CO_2-assisted metallization bath containing 0.2 vol.% of the surfactant and 40 ml of the Ni-P metallization solution. Total volume of the reaction cell is 50 ml, which gave 10 ml available for the CO_2 or 20 vol.% in volume fraction. Agitation of 400 rpm was used to ensure stability. The reaction temperature was 80 °C for both metallization method, and pressure of 15 MPa was used for the sc-CO_2-assisted metallization. The reaction time for the CONV metallization was 20 min. Three values of the reaction time for the sc-CO_2-assisted metallization were used, which were 20, 40, and 60 min. Surface of the Ni-P coated textile was characterized by an OM, SEM, and EDX equipped in the SEM to confirm conditions of the coating and measure film thickness of the Ni-P coating on the fiber.

For metallization of the non-electrical-conductive polymers, a catalyzation process is usually needed before the metallization process. Fig. 6.31 shows 3D images of overview of the textiles before and after the catalyzation processes. Generally, after the catalyzation process, for both the CONV catalyzation and sc-CO_2 catalyzation, no obvious difference on the outline of the textile structure was observed from the 3D images. The 3D images show the textiles were composed of several Nylon 6,6 fibers.

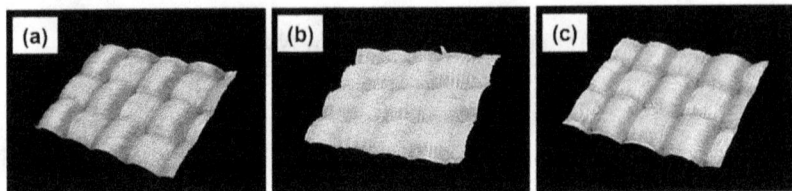

Fig. 6.31. 3D images of the textiles observed by the OM: (a) untreated Nylon 6,6 textiles, (b) catalyzed Nylon 6,6 textiles by the CONV catalyzation for 10 s, and (c) catalyzed Nylon 6,6 textiles by the sc-CO$_2$ catalyzation for 120 min.

Surface condition of the fibers was observed from OM images with higher magnification as shown in Figs. 6.32 and 6.33. Outline of each Nylon 6,6 fibers could be clearly observed in the top view OM images as shown in Figs. 6.32a and b. The strong acid used in the CONV catalyzation could cleave the chemical bonding of the polymers and roughen the surface to improve coverage and adherence of the metal coatings on the polymers. However, the fibers are often damaged. Nylon 6,6 polymer chains can be broken in the hydrolysis reaction with the acid used in the PdCl$_2$/SnCl$_2$ solution [31]. The white particulates are suggested to be the damaged polymers stayed on the fibers after the CONV catalyzation step, shown in Fig. 6.32d. In general, the rough surface is beneficial in metallization of the polymers to enhance the adherence. However, for applications in the wearable device, damaged and deformed fiber surface are fatal problems to maintain the reliability of the electronic device. On the other hand, the surface of the Nylon 6,6 textile treated by the sc-CO$_2$ catalyzation was very similar to that of the untreated textile, and no defects were observed as shown in Fig. 6.33. This result confirmed that the Nylon 6.6 fibers would not be damaged in the sc-CO$_2$ catalyzation.

After the metallization process, either by the CONV catalyzation or the sc-CO$_2$ catalyzation followed by the CONV metallization or sc-CO$_2$-assisted metallization, outline of the Nylon 6,6 fibers could still be identified from the 3D images shown in Fig. 6.34.

More details of the Ni-P coatings on the fibers could be observed from the SEM images shown in Fig. 6.35. For the Ni-P coatings deposited by the CONV catalyzation followed by the CONV metallization, the surface was not uniform and rough, and some nodules and cracks were observed as shown in Fig. 6.35a.

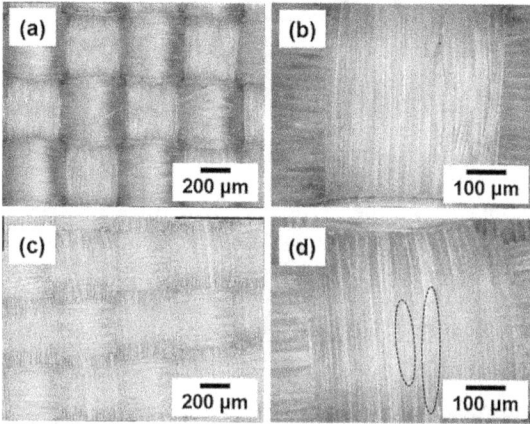

Fig. 6.32. OM images of the surfaces: untreated Nylon 6,6 textiles in (a) low
and (b) high magnification, and catalyzed Nylon 6,6 textiles by the CONV
catalyzation for 10 s in (c) low and (d) high magnification. The white rectangular
dash mark shows part of the damaged fiber and the black circular dash marks
shows the white particulates.

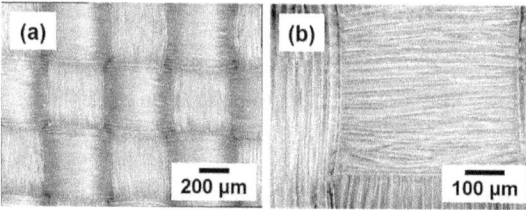

Fig. 6.33. OM images of the surfaces: Nylon 6,6 textiles treated by the sc-CO$_2$
catalyzation in (a) low and (b) high magnification.

The rough surface was suggested to be caused by the CONV catalyzation
and the hydrogen bubbles generated during the metallization reactions
provided in the following [32]:

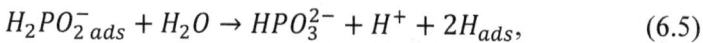

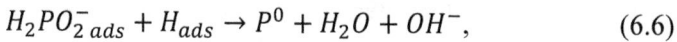

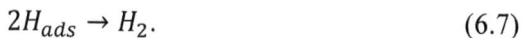

$$Ni^{2+}_{ads} + H_2PO^-_{2\,ads} + H_2O \rightarrow Ni^0 + 3H^+ + HPO^{2-}_3, \qquad (6.4)$$

$$H_2PO^-_{2\,ads} + H_2O \rightarrow HPO^{2-}_3 + H^+ + 2H_{ads}, \qquad (6.5)$$

$$H_2PO^-_{2\,ads} + H_{ads} \rightarrow P^0 + H_2O + OH^-, \qquad (6.6)$$

$$2H_{ads} \rightarrow H_2. \qquad (6.7)$$

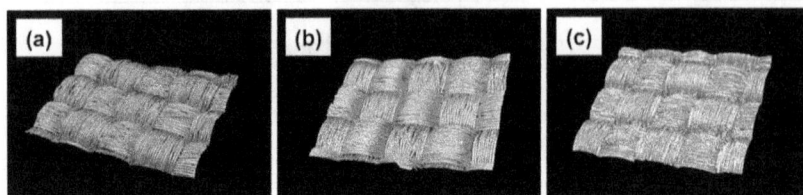

Fig. 6.34. 3D images of the Ni-P coated textiles observed by the OM: (a) The Ni-P coated textiles by the CONV catalyzation and the CONV metallization, (b) the Ni-P coated textiles by the sc-CO₂ catalyzation and the CONV metallization, and (c) the Ni-P coated textiles by the sc-CO₂ catalyzation and the sc-CO₂-assisted metallization.

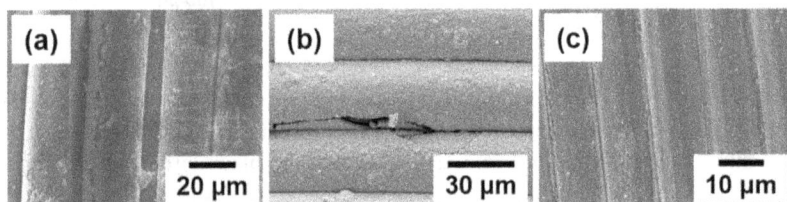

Fig. 6.35. SEM images of the Ni-P coated fibers: (a) The CONV catalyzation for 10 s and the CONV metallization for 1 min, (b) the sc-CO₂ catalyzation for 120 min and the CONV metallization for 20 min, and (c) the sc-CO₂ catalyzation for 120 min and the sc-CO₂-assisted metallization for 60 min.

Hydrogen bubbles adsorb on the surface to bring the defects. As shown in Figs. 6.35b and c, smooth surface and no pin-holes were observed on the Ni-P coating deposited by the sc-CO₂ catalyzation followed by the CONV metallization or the sc-CO₂-assisted metallization. However, the Ni-P coating deposited by the sc-CO₂ catalyzation followed by the CONV metallization had more peeled-off parts when compared with the coating by the sc-CO₂ catalyzation followed by the sc-CO₂-assisted metallization. The peeled-off parts demonstrated that the adherence of the Ni-P coating by the CONV metallization was low because of the insufficient anchoring effect of the metal coating on the fiber [33]. In contrast, the adherence of the Ni-P deposited by the sc-CO₂-assisted metallization was high because of the high anchoring effect given by impregnation of the Pd catalysts into the Nylon 6,6 fibers. The smooth surface of the Ni-P deposited by the sc-CO₂-assisted metallization was suggested to be caused by the sc-CO₂ electrolyte [34]. The electrolyte contains a continuous phase composed of the aqueous electrolyte and dispersed phases composed of sc-CO_2 and the surfactant. Surface tension of sc-CO_2 is very low, and CO_2 is nonpolar. Also, the dispersed phases would continuous bounce on the substrate surface during the metallization process, which then enhance removal of the adsorbed hydrogen gas bubbles from surface of the substrate to prevent generation of the defects [35, 36].

Cross-sectional SEM and EDX images of the Ni-P coating deposited on the Nylon fiber are shown in Fig. 6.36. Some peeled-off parts could be observed in the Ni-P coating deposited by the CONV catalyzation followed by the CONV metallization as indicated by the arrows in Fig. 6.36a. The average thickness of the Ni-P coating deposited by the CONV catalyzation (10 s) and the CONV metallization (1 min) was 7.03 μm. The coating thickness is thicker than the Ni-P coating deposited by the sc-CO_2 catalyzation (120 min) followed by the CONV metallization (10 min) and the sc-CO_2 catalyzation (120 min) followed by the sc-CO_2-assisted metallization (60 min), which were 5.28 and 1.06 μm, respectively. In general, the Ni-P deposition rate was lower when either or both the sc-CO_2 catalyzation and the sc-CO_2-assisted metallization were used. The low deposition rate for the sc-CO_2 catalyzation when compared with the CONV metallization is suggested to be a result of the difference in the crystallinity of the Pd catalyst after the catalyzation step. The CONV catalyzation is expected to give Pd nanocrystallines on the Nylon 6,6 surface. On the other hand, the sc-CO_2 catalyzation could impregnate the Pd catalyst deep into the Nylon 6,6 to improve the adherence but the crystallinity of the Pd is poor. The Pd impregnated into the Nylon 6,6 is either Pd crystals having size in the atomic scale or Pd ions in the form of the Pd organo-complex. The Pd organo-complex would be reduced to metallic Pd after immersion in the metallization solution since there are reducing agents in the metallization solution. The poor crystallinity of the Pd catalyst by the sc-CO_2 catalyzation caused the low efficiency of the Ni reduction reaction and the low deposition rate. Regarding the low deposition rate observed in the sc-CO_2-assisted metallization, it is suggested to be caused by a phenomenon often observed in applications of the emulsified electrolyte, which is named the "Periodic Plating Characteristic (PPC)". For the PPC, adsorption and desorption of the sc-CO_2 dispersed phases from surface of the substrate could retard the metal ion reduction reaction periodically [35]. However, the PPC might not be enough to explain the low deposition rate in the sc-CO_2-assisted metallization. Another explanation is related to dissociation of CO_2 into the aqueous metallization solution. The dissociation of CO_2 into the metallization solution would cause a decrease in the pH as indicated in Eq. (6.8) [37, 38].

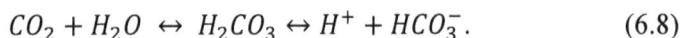

$$CO_2 + H_2O \leftrightarrow H_2CO_3 \leftrightarrow H^+ + HCO_3^-. \qquad (6.8)$$

When the pH of the metallization solution is low, it is expected to cause a decrease in the film growth rate [4]. Also, the surface smoothening effect is suggested to a result of the PPC [36]. The continuous adsorption and desorption of the sc-CO_2 dispersed phases to a fine convex part on the Ni-P coating can affect the nodule formation during the sc-CO_2-assisted metallization and eventually suppresses the growth of the nodules.

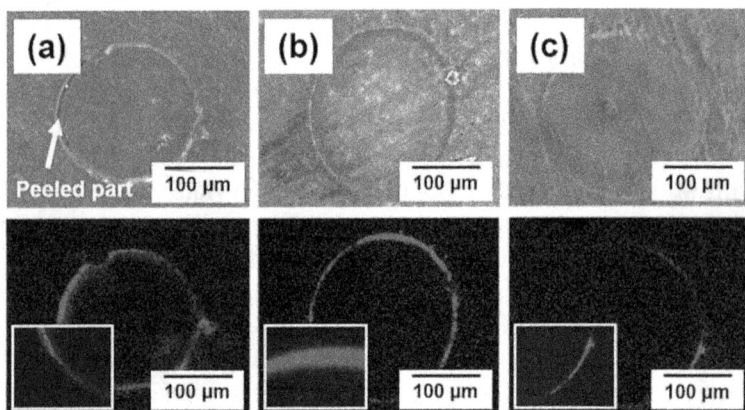

Fig. 6.36. Cross-sectional SEM and EDX images of the Ni-P coatings on the fiber: (a) the CONV catalyzation for 10 s and the CONV metallization for 1 min, (b) the sc-CO_2 catalyzation for 120 min and the CONV metallization for 20 min, (c) the sc-CO_2 catalyzation for 120 min and the sc-CO_2-assisted metallization for 60 min. Red parts are signals of Ni detected by the EDX. The white arrows indicate the peeled parts in (a). (For interpretation of the references to color in this figure legend, the reader is referred to the web version of this article.)

Fig. 6.37 shows the Ni-P coatings deposited by various lengths of the catalyzation and metallization times in the sc-CO_2 catalyzation and the sc-CO_2-assisted metallization, respectively. The amount of the deposited Ni-P was increased with an increased in either or both of the catalyzation time and the metallization time. The Pd organo-complex could dissolve into the sc-CO$_2$ and then impregnate into the Nylon 6,6 fibers. With a longer catalyzation time, the amount of the Pd catalyst into the fibers would be increased and caused an increase in the amount of the Ni-P deposited. The increase in the Ni-P coverage with an increase in the metallization time during the sc-CO$_2$-assisted metallization is expected since a longer metallization time gives a higher amount of the reduced Ni-P.

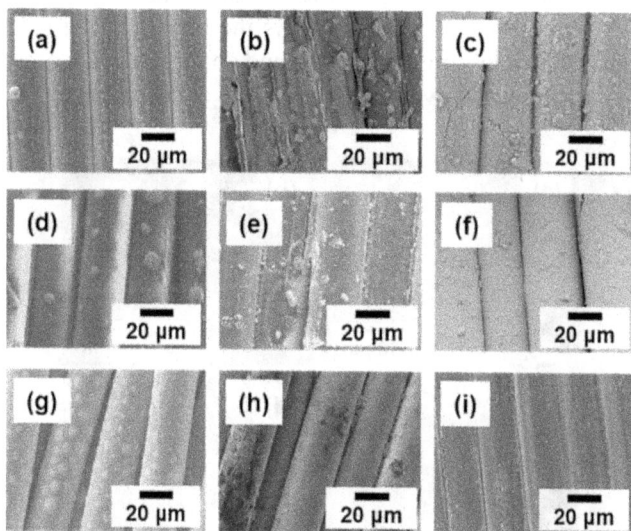

Fig. 6.37. Surfaces of the Ni-P coatings with various time in the sc-CO$_2$ catalyzation and the sc-CO$_2$-assisted metallization: (a) sc-CO$_2$ catalyzation for 20 min and sc-CO$_2$-assisted metallization for 20 min, (b) sc-CO$_2$ catalyzation for 20 min and sc-CO$_2$-assisted metallization for 40 min, (c) sc-CO$_2$ catalyzation for 20 min and sc-CO$_2$-assisted metallization for 60 min, (d) sc-CO$_2$ catalyzation for 60 min and sc-CO$_2$-assisted metallization for 20 min, (e) sc-CO$_2$ catalyzation for 60 min and sc-CO$_2$-assisted metallization for 40 min, (f) sc-CO$_2$ catalyzation for 60 min and sc-CO$_2$-assisted metallization for 60 min, (g) sc-CO$_2$ catalyzation for 120 min and sc-CO$_2$-assisted metallization for 20 min, (h) sc-CO$_2$ catalyzation for 120 min and sc-CO$_2$-assisted metallization for 40 min, (i) sc-CO$_2$ catalyzation for 120 min and sc-CO$_2$-assisted metallization for 60 min.

In order to confirm the deposition mechanism, EDX analysis of the fiber surfaces after the sc-CO$_2$-assisted metallization was conducted as shown in Fig. 6.38. When 120 min of the sc-CO$_2$ catalyzation and 20 min of the sc-CO$_2$-assisted metallization were used, nodule-like morphology was observed, and the nodule was confirmed to be mostly composed of Ni as shown in Fig. 6.38a. As the metallization time was increased to 40 min, an increase in size of the nodules was observed. Complete coverage of the Nylon 6,6 fibers with the Ni-P coatings was obtained when the metallization time was increased to 60 min as shown in Fig. 6.38c. The results obtained from the EDX analysis showed that the Ni reduction would be initiated at specific parts on the fiber and grow into hemispherical-shaped particles having size in several micro-meters. At this stage, the hemispherical particles are not conducted to each other, and the distribution is quite uniform. As the metallization proceed, the hemispherical particles gradually grow and connect to each other to form continuous coatings and eventually cover the entire surface.

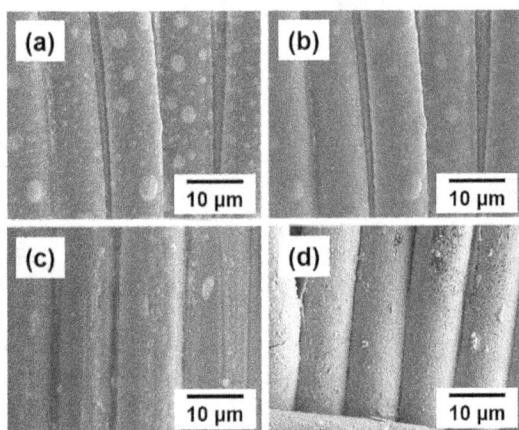

Fig. 6.38. SEM and EDX mapping images of the Ni-P deposited on the fibers by (a) and (b) the sc-CO_2 catalyzation for 120 min and the sc-CO_2-assisted metallization for 20 min, (c) the sc-CO_2 catalyzation for 120 min and the sc-CO_2-assisted metallization for 40 min, and (d) the sc-CO_2 catalyzation for 120 min and the sc-CO_2-assisted metallization for 60 min. Red parts are the Ni content and green parts are the P content detected by EDX. (For interpretation of the references to color in this figure legend, the reader is referred to the web version of this article.)

6.2.3. Nickel-Phosphorus Metallization on Silk

As-received silk substrates with dimensions of 2 cm × 4 cm were used in this study. Identical CONV catalyzation solution was used for CONV catalyzation on silk. Same high pressure apparatus were used for sc-CO_2 catalyzation. 25 mg of the catalyst was used with respect to the 50 ml reaction cell to maintain the catalyst concentration at the saturation point throughout the process. No pretreatments were carried out before the catalyzation step. The catalyzation was executed at 80 ± 1 °C and 15 ± 0.1 MPa with agitation for 2 hours. Identical Ni-P metallization solution was used in this section. The metallization was executed at an isothermal environment controlled at 70 ± 1 °C under ambient pressure with agitation. Various metallization times (t = 0.5, 1, 2, 3, 4, 5, 10, and 20 min) were performed to examine the relationship between the metallization time and the metallization characteristics. No post treatment was conducted after the metallization. Surface morphologies, thickness, compositions, and phases were observed by an OM, a SEM, an ImageJ software (National Institutes of Health, U.S.A.), an EDX, and a XRD. The electrical resistance was estimated by a four-point probe using specimens having dimensions of 1 cm $\times$ 0.5 cm (length$\times$width) at room temperature. Twenty electrical resistance measurements were conducted for each specimen. The adhesive test was conducted by sticking a piece of 3M tape (3M adhesive tape #810, 3M, USA) onto surface of the Ni-P/silk firmly and then peeling-off the 3M tape from the surface. Dimensions of the Ni-P/silk composite material evaluated were 1 cm $\times$ 0.5 cm. The electrical resistance was measured before and after the adhesive tests to check the durability and firmness of the metallized layer. Corrosion resistance was measured with a 3.5 wt.% NaCl solution at room temperature. A piece of the Ni-P/silk composite material with 4 min of the metallization time was used as the working electrode. A piece of Pt plate was used as the counter electrode, and Ag/AgCl was used as the reference electrode. The Pt counter electrode was cleaned ultrasonically in ethanol and pure water for 1 min, respectively, before each measurement. The corrosion test was conducted by scanning the potential from -0.7 to + 0.3 V at 1 mV/s scan rate. Prior to the corrosion test, the electrodes were immersed in the 3.5 wt.% NaCl solution for 25 min to certify and establish a stable open circuit potential (OCP). Surface area of the working electrode, that is the contact area between the Ni-P layer and the electrolyte, was estimated to be 9.5 cm^2, which was calculated by estimating every bundles of the Ni-P metallized silk in the specimen. Potentiodynamic polarization curves were obtained using a potentiostat/galvanostat (1287A, Solartron Analytical, UK).

Figs. 6.39a-c show the as-received silk, CONV catalyzed silk, and sc-CO_2 catalyzed silk, respectively. Transparent thread bundles shown in Fig. 6.39a indicate the silk substrate. Obvious damages (circled by dash lines), which were deteriorated by the acidic HCl solution, can be observed in Fig. 6.39b. The silk loses its structure after the CONV catalyzation treatment showing a rough surface. The light yellow flakes observed in Fig. 6.39c correspond to the Pd(acac)$_2$ catalyst. With the help of the sc-CO$_2$, the catalyst was successfully embedded into the silk substrate without damaging the silk structure (Fig. 6.39c).

Fig. 6.39. OM images of (a) the non-catalyzed silk, (b) CONV catalyzed silk, and (c) sc-CO$_2$ catalyzed silk.

X-ray diffraction patterns of the as-received silk, the sc-CO$_2$ catalyzed silk, the catalyzed/reducing agent treated silk, and the catalyzed/Ni-P electrolyte treated silk are shown in Fig. 6.40. In Fig. 6.40a, a board peak at around $2\theta = 20°$ indicates amorphous structure of the silk. The characteristic peaks in the Fig. 6.40b correspond to the Pd(acac)$_2$ well indicating the catalysts were successfully settled on the substrate. Due to the high intensity from Pd(acac)$_2$ catalyst, no silk characteristic peaks were observed in Fig. 6.40b. Fig. 6.40c shows the reduced catalysts on the silk substrate, Pd(acac)$_2$ catalysts were reduced to metallic palladium after immersing the catalyzed silk in a solution containing the reducing agent for 15 min at 70 °C. According to Y. S. Cheng et al. [39] the reduction reaction of Pd(acac)$_2$ is believed to be the following reaction:

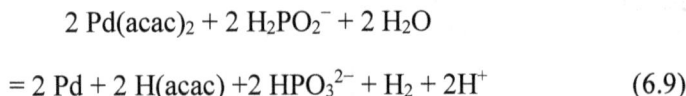

$$2\ Pd(acac)_2 + 2\ H_2PO_2^- + 2\ H_2O$$

$$= 2\ Pd + 2\ H(acac) + 2\ HPO_3^{2-} + H_2 + 2H^+ \qquad (6.9)$$

After the reduction, the catalyst-contained silk surface became active and allowed initiation of the Ni-P metallization [40]. Four diffraction peaks, locate at $2\theta = 40.1°$, $46.6°$, $68.1°$, and $82.1°$ can be indexed to (111), (200), (220), and (311) planes of the FCC structure for palladium (JCPDS #89-4897). In Fig. 6.40d, a broad peak locates at $2\theta = 20°$ can be indexed to the silk substrate, while the other broad peak shows around 40-50 degree representing the amorphous Ni-P metallized layer. No Pd or Pd(acac)$_2$ characteristic peaks can be observed after the metallization indicating that the small amount of Pd contained precursors were thoroughly covered by the deposited Ni-P.

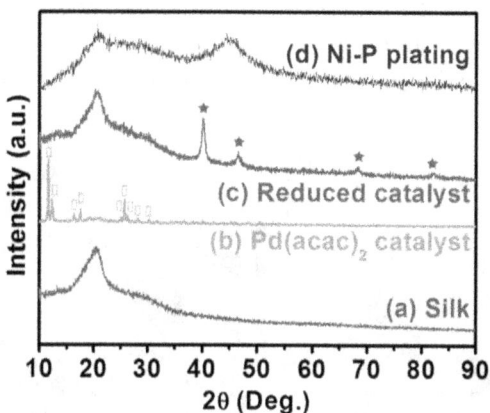

Fig. 6.40. X-ray diffractions of (a) non-catalyzed pure silk, (b) sc-CO$_2$ catalyzed silk, (c) sc-CO$_2$ catalyzed/reduced silk, and (d) Ni-P metallized silk respectively.

Morphologies of the metallized silks with various metallization times are shown in Fig. 6.41. Figs. 6.41a and b show the OM and SEM images of the Ni-P metallized silk deposited for 30 s and 1 min, respectively. As shown in Fig. 6.41a, small amounts of the catalysts were still unreduced (circled) and part of the silk was still un-metallized showing the transparent color after 30 s of the metallization time. No SEM image is shown here due to the poor conductivity of the specimen. Similarly, in Fig. 6.41b, the bright color in the inserted SEM image indicates the silk surfaces are still not fully covered by the Ni-P layer (pointed by arrows), and the Ni-P layer is still not continuous resulting a high electrical resistance state (results shown in the following section). Results obtained in this study are classified into three stages. The specimens not fully covered by the Ni-P are classified into the first stage, which are samples metallized at 30 s and 1 min. On the other hand, in Fig. 6.41c, the Ni-P layer shows a continuous and complete coverage at 3 min of the metallization time. The silk is fully covered when the metallization time was extended to 3 min. The silk substrate metallized for 4 min is shown in Fig. 6.41d demonstrates the full coverage and a consecutive growth of

211

the Ni-P layer thickness. The samples metallized for 3 and 4 min of the metallization time both having a complete coverage and smooth surface are classified into the second stage. Beyond 5 min of the metallization time, rough surface was formed, which resulted deteriorated physical properties (discussed in the later section). Meanwhile, redundant Ni-P clusters (pointed by arrows) accumulated on the surface with an increase in the metallization time (Figs. 6.41e and f). The specimens with the metallization time longer than 5 min are classified into the third stage.

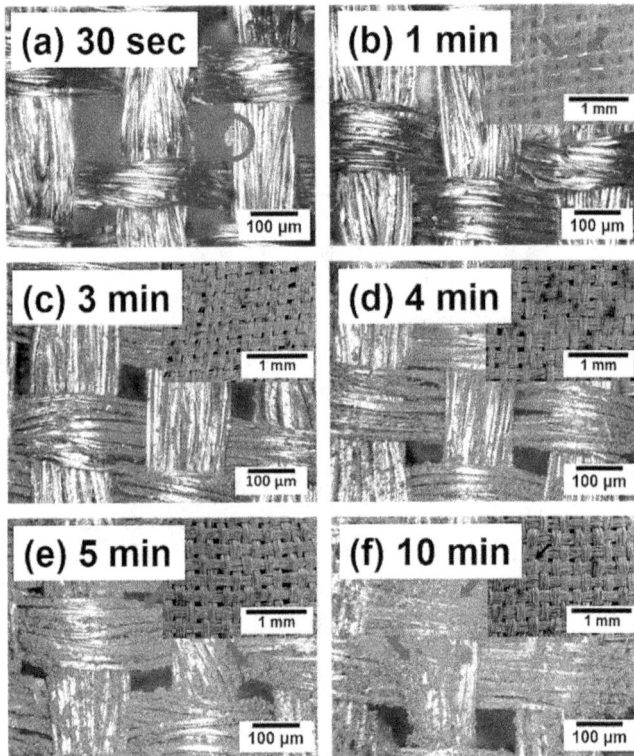

Fig. 6.41. OM images and the inserted SEM images of the Ni-P coated silk composites prepared by (a) 30 sec, (b) 1 min, (c) 3 min, (d) 4 min, (e) 5 min, and (f) 10 min of the metallization time.

Elemental mapping results from the 4 min metallization time are presented in Fig. 6.42a. All compositions of Ni-P layer in various metallization times were revealed in Table 6.1. The cross section of the Ni-P metallized silk for 4 min is shown in Fig. 6.42b.

Table 6.1. Ni-P compositions in various metallization times.

Comp.	1 min	2 min	3 min	4 min	5 min	10 min	20 min
Ni (at.%)	83.8	83.8	83.9	82.2	82.1	82.5	82.4
P (at.%)	16.2	16.2	16.1	17.8	17.9	17.5	17.6

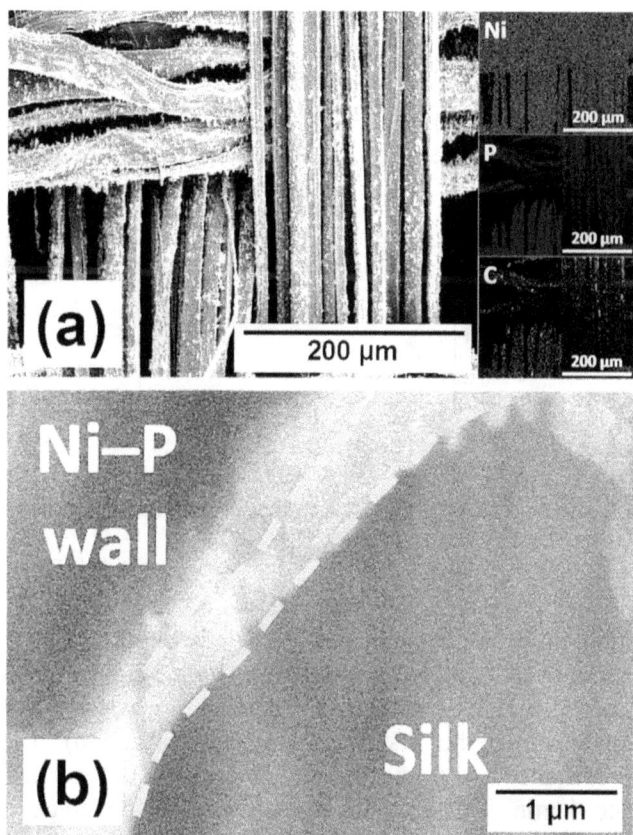

Fig. 6.42. (a) Elemental mapping of the Ni-P/silk composite prepared by 4 min
of the metallization time and (b) cross section of Ni-P metallized
for 4 min on silk.

The metallization time dependence of the Ni-P layer thickness is
presented in Fig. 6.43. Positive correlation in the metallization time and
the thickness is demonstrated. Thickness of the Ni-P layer with the

longest metallization time (20 min) was 0.7 µm. A decelerated growth rate can be observed when the metallization time is longer than 5 min, which can be attributed to the lowered Ni ion and reducing agent concentrations in the bulk metallization solution. Ni ions and reducing agent in the solution were consumed rapidly in the metallization step. At the beginning of the metallization, the concentrations were high and gave a Ni-P layer growth rate at approximately 0.08 µm/min in the initial 5 min of the metallization time. On the other hand, the Ni ion and the reducing agent concentrations continued to drop and caused the growth rate to slow down to approximately 0.005 µm/min after 5 min of the metallization time.

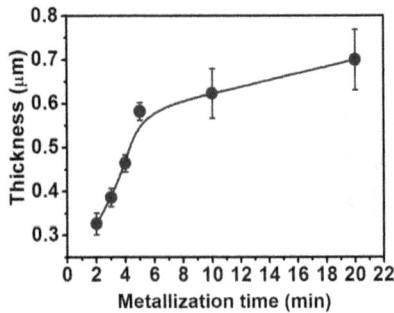

Fig. 6.43. The Ni-P layer thickness as a function of the metallization time.

Fig. 6.44 illustrates the electrical resistance as a function of the metallization time before and after the adhesive tests. The Ni-P/silk composites showed high electrical resistance since the conductive metal layer was discontinuous when the metallization time was less than 3 min, which is referred as the first stage (Figs. 6.41a and b). The lowest electrical resistance at 1.02 was achieved as the metallization time was extended to 4 min. This result is in line with the coverage shown in Fig. 6.41 indicating that the coverage is one of the critical factors affecting the electrical property. Time duration between the beginning of the full coverage and the time required to reach the minimum resistance is corresponded as the second stage. Metallization time beyond 5 min is classified into the third stage, which the electrical resistance is offset by the roughened surface. In the third stage (Figs. 6.41e and f), the growth of rough and loose particles on the substrate surface would bring unnecessary Joule heating and more electron scattering, which are the main reasons for the deteriorated electrical resistance [41].

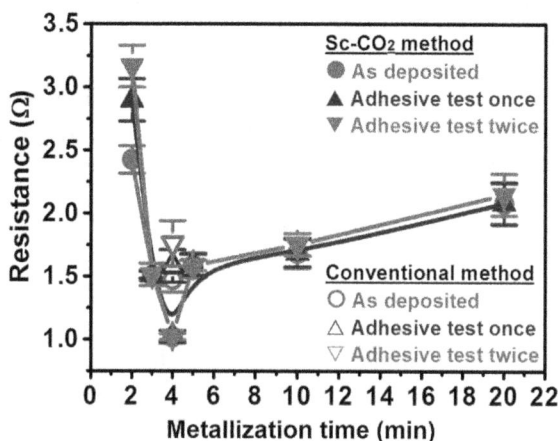

Fig. 6.44. Electrical resistance curves of the Ni-P/silk composites before and after the adhesive test.

In addition, a direct evidence indicating the increase of the roughness with respect to the metallization time is revealed in Fig. 6.45. Fig. 6.45a shows surface morphology of the Ni-P/silk composite with 1 min of the metallization time. The surface covered by the Ni-P layer is smooth and charging effect is observed at the region which was not covered by the Ni-P layer. On the other hand, Figs. 6.45b and c demonstrate the surface roughness increased when the metallization time is increased to 2 and 5 min. A proportional relationship between the electrical resistance and the roughness has been indicated in a previous study [42]. Standard deviation of the electrical resistance decreases with the metallization time indicating that the inhomogeneity can be mitigated as the metallization time extends in the first two stages. However, the standard deviation increases after the second stage due to the increased surface roughness. A reliable electrical resistance cannot be obtained owing to the poor conductivity when the metallization time is less than 2 min. Specimen prepared by the CONV catalyzation and metallization at 4 min is also shown in Fig. 6.44. Specimens fabricated by the CONV catalyzation show higher electrical resistance than those catalyzed by sc-CO_2 method due to the deteriorated substrate surface. Rough substrate surface results in the coarse deposition of Ni-P and finally leads to the high electrical resistance. Moreover, since the catalysts are only attach to the substrate surface, electrical resistance changed significantly after adhesive tests.

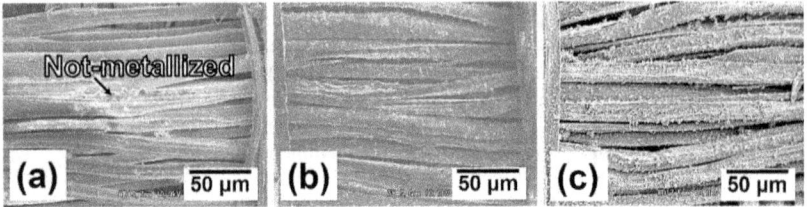

Fig. 6.45. SEM images showing surface condition of the Ni-P/silk composite prepared by (a) 1 min, (b) 2 min, and (c) 5 min of the metallization time.

A summary of the influence of the morphology on the electrical resistance in different stages is illustrated in Fig. 6.46. In the first stage, the Ni-P layer is non-continuous resulting in poor electrical conductivity. When the metallization time is extended to the second stage, a continuous Ni-P layer is constructed, in other words, the silk is fully covered. A full coverage, a smooth surface, and a sufficient Ni-P layer thickness result in the lowest electrical resistance. In the last stage, the electrical conductivity is offset by the increased roughness. The electrical resistances after the adhesive tests are shown in Fig. 6.44.

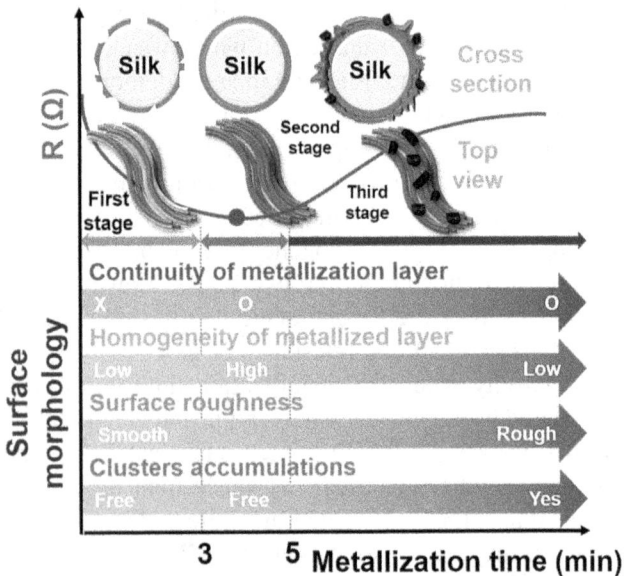

Fig. 6.46. A summary of the influence of the morphology on the electrical resistance.

The electrical resistances persist in the second and third stages even after twice of the adhesive test. In addition, there is no distinct change among these three measurements. However, the adhesive tests show serious impact on the samples in the first stage. Since the joint of the un-metallized silk surface and the Ni-P layer can be the weak points of the composite materials, thus, it was destructed more severely than the Ni-P fully covered samples. The Ni-P/silk composite metallized with the optimized time shows perseverance in the adhesive tests indicating its feasibility in the practice of wearable devices even under adverse conditions.

Corrosion behaviors of the Ni-P/silk composites in 3.5 wt.% NaCl are shown in Fig. 6.47. Fig. 6.47a shows the Tafel plots of the as-deposited Ni-P/silk composite and the Ni-P/silk composite after the adhesive tests. The Ni-P/silk composite prepared by 4 min of the metallization time was chosen for the corrosion resistance evaluations. No significant difference was found among the three curves indicating excellent reliability of the Ni-P layer on the silk surface even after twice of the adhesive test. The corrosion potential (E_{corr}) and the corrosion current density (I_{corr}) are summarized in Table 6.2. The I_{corr} are comparable to those of Ni-P layer on metal substrate reported in other studies [43-45], which indicate the high corrosion resistance of the Ni-P layer retained even when it is deposited on a non-metallic substrate. Moreover, Fig. 6.47b shows the morphology after twice-adhesive tests and the corrosion test, no obvious corrosion attack can be observed after the testing. The low corrosion rate in a solution simulating the human sweat before and after the adhesive test demonstrates the Ni-P/silk is practical in wearable devices.

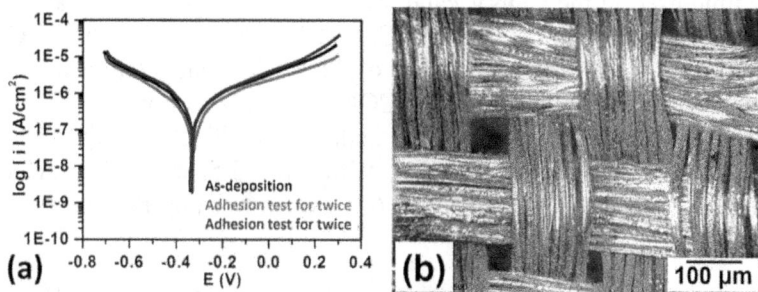

Fig. 6.47. (a) Tafel plots of the Ni-P/silk composite prepared with 4 min of the metallization time before and after the adhesive test, and (b) an OM image of the Ni-P /silk composite after twice-adhesive tests and the corrosion test.

Table 6.2. Corrosion potentials and corrosion current densities of the deposited Ni-P/silk composite material.

Samples	E_{corr} (V)	I_{corr} (A/cm^2)
(a) As-deposited	-0.32	5.89×10^{-7}
(b) Adhesive test for once	-0.33	6.00×10^{-7}
(c) Adhesive test for twice	-0.32	5.93×10^{-7}

6.3. Platinum Metallization

6.3.1. Platinum Metallization on Nylon 6,6

The high pressure apparatus and reaction chamber are identical to the previous sections. The reaction chamber was placed in a temperature-controlled air bath to control the experimental temperature. The Nylon 6,6 textile was positioned in the reaction chamber using a plastic-coated platinum wire. Same CONV catalyzation solution was used in this section. Pt(acac)$_2$ (99.98 %, Sigma-Aldrich, USA) was used as the source of Pt in the sc-CO$_2$ catalyzation. The Pt metallization bath was commercially available electrolyte purchased from MATEX Japan, and it was consisted of MATEX PLATINUM basic bath (500 mL/1 L), MATEX PLATINUM reduction agent (10 mL/1 L), NH$_3$ solution (50 mL/1 L), and ion-exchanged water (460 mL/1 L). Platinum concentration in the metallization bath was 2 g/L. The pH of this bath was 12. The textile used in this study was made from Nylon 6,6 fibers.

Image of the untreated Nylon 6,6 is shown in Fig. 6.48a. No pretreatment was conducted. Two types of catalyzation procedures were performed. The temperature used in the CONV catalyzation was room temperature. The time used in the CONV catalyzation was 10 s. The temperature of the sc-CO$_2$ catalyzation was 80 °C, and the pressure for the sc-CO$_2$ catalyzation was 15 MPa. The time of the sc-CO$_2$ catalyzation was 2 h. After the catalyzation process, the CONV metallization was performed. The catalyzed textile was immersed in the Pt metallization bath at 60 °C. Two immersion times were used, which are 3.5 and 4 h. Surfaces morphology of the Pt coated Nylon 6,6 textiles was observed by an OM and a SEM.

Images of the Nylon 6,6 textiles treated by the CONV catalyzation and 3.5 and 4 h of the CONV metallization are shown in Figs. 6.48b and c, respectively. By comparing the appearance of the CONV catalyzation and CONV metallization treated Nylon 6,6 textiles with the untreated

Nylon 6,6 textile shown in Fig. 6.48a, surface of the CONV catalyzation treated Nylon 6,6 fibers was rough and a few white particulates were observed under the OM. The results were suggested to be caused by the acid used in the CONV catalyzation step. Rough surface is usually beneficial in metallization of polymers. However, surface of the CONV catalyzation treated Nylon 6,6 fibers did not show obvious appearance of any metallic coating after 3.5 h of the CONV metallization process. When the Pt metallization time was increased to 4 h, the appearance was still similar to that of the 3.5 h case.

The sc-CO_2 catalyzation and CONV metallization treated Nylon 6,6 textiles are shown in Fig. 6.49. After the CONV metallization for 3.5 h, shown in Fig. 6.49a, surface of the Nylon 6,6 fibers was found to be covered by bright Pt coating. The Pt coatings seemed to be very thin in the OM observation. Therefore, a longer metallization time of 4 h was used to increase the coating thickness. As shown in Fig. 6.49b, much thicker Pt coating was observed on the Nylon 6,6 fibers. Although the Pt coating became much thicker after 4 h of the metallization, the outline of the fibers could still be seen.

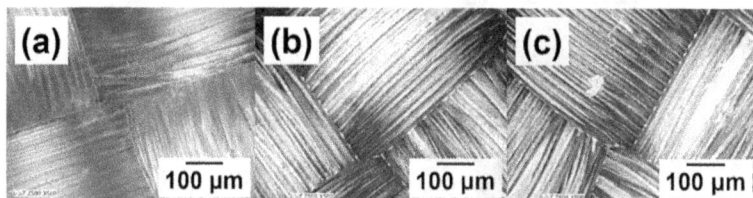

Fig. 6.48. OM images of the (a) untreated Nylon 6,6 textile and Nylon 6,6 textiles treated by (b) CONV catalyzation and CONV metallization for 3.5 h and (c) CONV catalyzation and CONV metallization for 4 h.

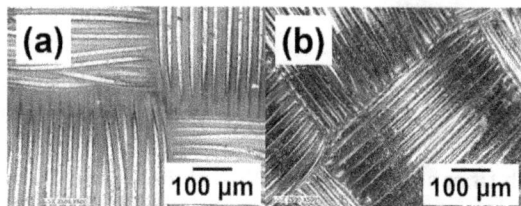

Fig. 6.49. OM images of the Nylon 6,6 textiles treated by (a) sc-CO_2 catalyzation and CONV metallization for 3.5 h and (b) sc-CO_2 catalyzation and CONV metallization for 4 h.

219

More details of the Pt coatings could be observed from the SEM images in Fig. 6.50. For, the CONV catalyzation and CONV metallization treated Nylon 6,6 textiles, the surfaces were rough, and it was difficult to identify outline of the fibers because boundary between each fibers was not clear, shown in Figs. 6.50a and b. Also, there was not much difference when the CONV metallization was increased from 3.5 to 4 h. These results were in agreement with the surface morphology observed by the OM and confirmed that the CONV catalyzation process was not appropriate for metallization of the Nylon 6,6 textiles. This is because; damage of the Nylon 6,6 and deformation of the fibers occurred in the CONV catalyzation step. The acid used in the $PdCl_2/SnCl_2$ solution can react with the Nylon 6,6 polymer chains. Then, subsequent disruption of the hydrogen bonds between the polymer chains occurred [31].

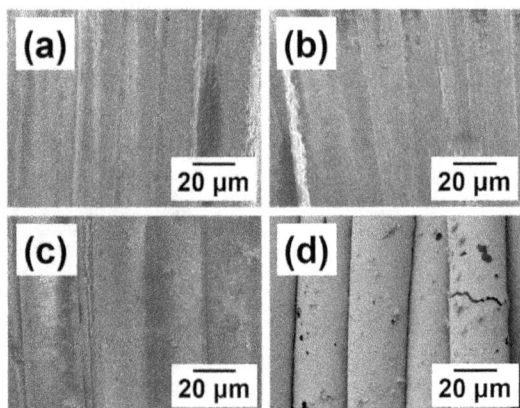

Fig. 6.50. SEM images of surface of the Pt coatings by (a) CONV catalyzation and CONV metallization for 3.5 h, (b) CONV catalyzation and CONV metallization for 4 h, (c) sc-CO_2 catalyzation and CONV metallization for 3.5 h, (d) sc-CO_2 catalyzation and CONV metallization.

In Fig. 6.50c, the SEM image shows that the sc-CO_2 catalyzation and CONV metallization for 3.5 h could produce pinhole-free and smooth Pt coatings on the Nylon 6,6 fibers. Furthermore, the Pt coating on the surface of each fiber still showed outline of the fibers. No peeled-off part was observed. These results demonstrate that the sc-CO_2 catalyzation was an excellent catalyzation process to improve the adherence of the metal coatings on the textile. The adherence was improved by the anchoring effect, which metal roots were formed under surface of the polymers to anchor the metal coatings [6]. The metal roots were formed by impregnation of the Pt catalysts into the polymer by the sc-CO_2. The solubility of Pt(acac)$_2$ pure sc-CO_2 is high, and sc-CO_2 can penetrate into the structures of polymers because of the non-polar and zero surface tension properties [28]. So, the catalyst could be transferred into the polymer structure to form catalyst seeds in the polymer. Then, reduction of the metal ions could be initiated at the catalyst seeds to form metal roots to anchor the metal coatings. Fig. 6.50d shows the Pt coating by the sc-CO_2 catalyzation and CONV metallization of 4 h. With a longer metallization time, the Pt coating thickness is expected to be higher. The result showed that the outline of the fibers was still obvious with a thicker Pt coating, but some cracks were observed. The cracks were suggested to be mostly caused by the differences in elastic constant between Pt and Nylon 6,6. A longer metallization time could also lead to a larger internal stress in the metallized film, which can lead to the cracks or peel-offs.

6.3.2. Platinum Metallization on Silk

6.3.2.1. Pt(acac)$_2$ Sc-CO_2 Catalyzation and CONV Metallization

One piece of silk textiles with 2 cm × 4 cm in width and length was hooked on the ceiling of a 50 ml stainless reaction cell. Sc-CO_2 together with Pt(acac)$_2$ catalyst were introduced into the reaction cell afterward. One piece of the silk incorporating with 25 mg of the catalyst was used to guarantee the catalyst is in saturation concentration. Identical high pressure apparatus was utilized in this section. The catalyzation was executed at 80 ± 1 °C under 15 ± 0.1 MPa for 2 h with agitation. No post treatments were carried out after the catalyzation step. The Pt-metallization electrolyte was the same as the previous section. The metallization was executed at 70 ± 1 °C in water bath under ambient pressure with agitation. Interrelationship between the metallization time lengths (100, 120, 140, 160, 180, and 200 min) and the characteristics of the Pt metallized silk were examined. No post treatments were conducted after the metallization.

Surface morphology of the metallized silk textiles was observed by an OM and a SEM. The Pt layer thicknesses were estimated by observing cross-sections of the silk-Pt composites by the SEM and Image-J software (National Institutes of Health, USA). The crystal structure and phases were identified by an XRD. In addition, the composition and coverage of the Pt metallized layer were measured by EDX. 1 cm × 1 cm in width and length of the silk-Pt composite was immersed in a SBF at 37 °C for 3 months to evaluate the ion releasing rate. 15 ml of the SBF was drawn out from the tube and centrifugated (Z206A, HERMLE Labortechnik, Germany) for 2 h to remove undesired particles. The ion releasing rate was determined by measuring the Pt ion concentration in the SBF by inductively coupled plasma atomic emission spectroscopy (ICP-AES; ICPS-8100, Shimadzu Co. Ltd., Japan). Samples with a sufficient metallization time were chosen for the ion releasing rate evaluation to ensure the silk surface was fully covered by Pt. Electrical resistance was measured at room temperature by a four-point probe with sample dimensions at 1 cm × 0.5 cm (length×width). Ten detections of ICP-AES and electrical resistance were conducted within one sample for appropriate reliance and setting up the standard deviation.

Figs. 6.51a and b shows the uncatalyzed silk and the silk textile after the catalyzation, respectively. The light yellow flakes correspond to the Pt(acac)$_2$ catalyst, and the transparent thread bundles represent the silk. By means of conducting the catalyzation with sc-CO$_2$, the catalysts can be distributed into the silk textile without damaging the silk structure. Although some of the catalysts aggregate in a small range, distribution of the catalysts in the substrate is nearly homogeneous.

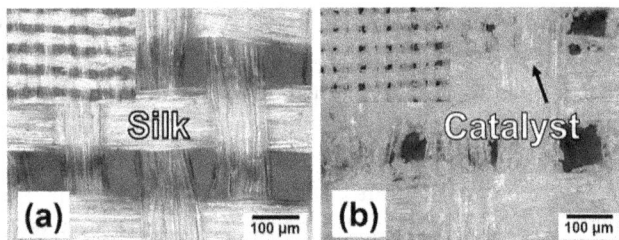

Fig. 6.51. OM images of (a) the non-catalyzed pure silk and (b) the catalyzed silk.

Fig. 6.52 demonstrates X-ray diffraction patterns of the (a) pure silk, (b) catalyzed silk, (c) silk with reduced catalyst, and (d) metallized silk, respectively. The diffraction pattern of the pure silk in Fig. 6.52a shows the amorphous structure. On the other hand, all the diffraction peaks in Fig. 6.52b correspond well to the Pt(acac)$_2$. In Figs. 6.52c and d, five diffraction peaks located at $2\theta = 40.2, 46.7, 67.9, 81.7,$ and $86.1°$ can be indexed to (111), (200), (220), (311), and (222) planes of the FCC structure for Pt (JCPDS #87-0647). A broad peak located at around $2\theta = 20°$ in Fig. 6.52c is the contribution from the silk textile. No characteristic peak of unreduced catalyst or other impurity phases were observed within the detection limit after the Pt metallization step.

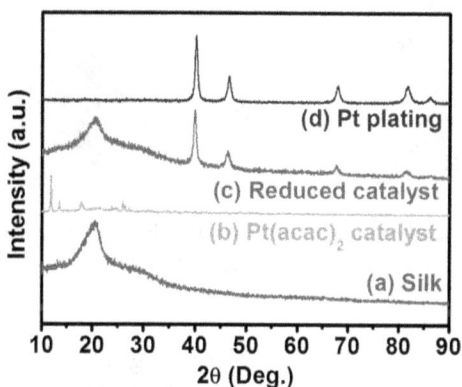

Fig. 6.52. X-ray diffraction patterns of (a) the non-catalyzed pure silk, (b) the catalyzed silk, (c) the silk with reduced catalyst, and (d) the Pt metallized silk.

The coverage tendency at 120, 160, and 200 min of the Pt metallization time durations is shown in Fig. 6.53. The bright phase corresponds to the silk, and the gray phase indicates the Pt metallization layer. In Fig. 6.53a, the Pt metallization layer is still non-continuous, which is referred as the first stage. While the metallization time is extended to 160 min in Fig. 6.53b, the silk textile is fully metallized. Similar morphology can be observed when the metallization time is extended to 140 min (not shown here). The silk-Pt composites with full coverage and smooth metallized surface are classified into the second stage. When the metallization times are extended to 180 and 200 min (third stage), complete coverage can be constructed along with Pt cluster accumulations (indicated by black arrows) and rough surface (Fig. 6.53c). Fig. 6.53 shows that coverage of the Pt metallized layer increases as a function of the metallization time length. However, rough morphology and undesired Pt clusters deposited on the surface are observed as shown in Fig. 6.53c, pointed by the black arrows. The roughened surface would inevitably deteriorate the electrical property (discussed below) in the third stage. On the other hand, the composition analysis results verified the as-deposited Pt layer to be pure Pt with trace impurity, which is in good agreement with the result from the X-ray diffraction pattern. Electrical conductivity of the Pt metallized silk is directly related to the coverage, which will be discussed in the following section.

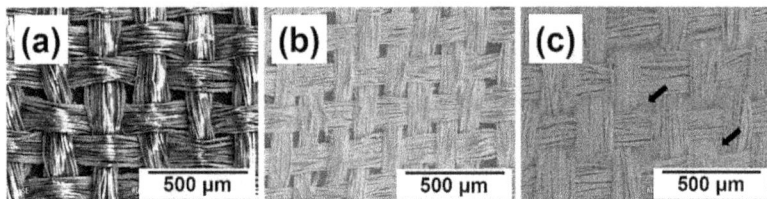

Fig. 6.53. Pt metallization trend shown by SEM images at (a) 120 min, (b) 160 min, and (c) 200 min of the metallization time, respectively.

In the same way, the metallized Pt thickness grows with the metallization time length. The dependence between the metallization time duration and the thickness is presented in Fig. 6.54. The maximum Pt thickness deposited with the longest metallization time at 200 min is 0.91 μm. The standard deviation decreases with the metallization time in the first two stages suggesting better deposit homogeneity. On the contrary, the standard deviation increases as a function of the metallization time in the third stage indicating that the homogeneity decreases as the metallization time prolongs. This can be attributed to the increased surface roughness. Reliable thickness cannot be measured when the metallization time was less than 140 min due to poor conductivity.

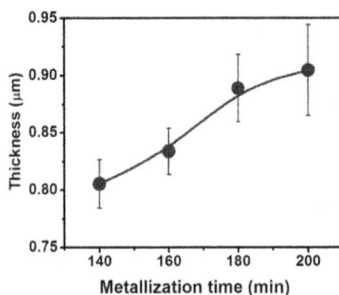

Fig. 6.54. Pt metallization layer thickness as a function of the metallization time.

Fig. 6.55 demonstrates the electrical resistance as a function of the metallization time period. The lowest electrical resistance at 86 mΩ is obtained as the metallization time is extended to 160 min with 100 % Pt coverage and smooth surface in the second stage. This result is in line with the coverage shown in Fig. 6.53 indicating that the coverage acts as a critical factor to the electrical property. On the contrary, the silk-Pt composite still remains high electrical resistance when the metallization time is less than 140 min (Fig. 6.53a) since the Pt layer is discontinuous in the first stage. The electrical resistances rise up in the third stage at 180 and 200 min of the metallization time durations, which can be attributed to the rough surface. As shown in Fig. 6.53c, the high surface roughness could bring unnecessary Joule heating and electron scattering, which are the main reasons for the deteriorated electrical resistance [41]. Therefore, 160 min is the optimized metallization time for this full Pt-contained electroless plating process. Similarly, the standard deviation decreases with the metallization time duration in the first two stages indicating that the inhomogeneity can be mitigated as the metallization time extends. On the contrary, the standard deviation increases in the third stage due to the high surface roughness. Reliable electrical resistance cannot be obtained for the samples with the metallization time less than 120 min owing to the poor conductivity.

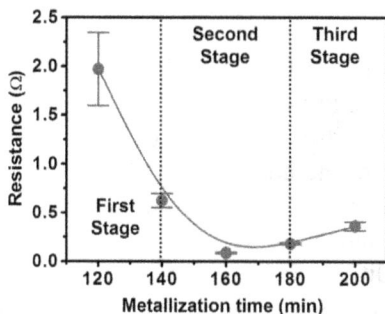

Fig. 6.55. Electrical resistance of the composite as a function of the metallization time.

A summary of the relationship between the morphology and the electrical resistance in different stages is shown in Fig. 6.56. In the first stage, parts of the silk textile are still un-metallized resulting in a high electrical resistance. When the metallization time is extended to the second stage, the Pt metallization layer becomes continuous along with smooth surface leading to the lowest electrical resistance. However, in the third stage, when the metallization time is beyond the optimized time, the electrical resistance is deteriorated due to the rough surface.

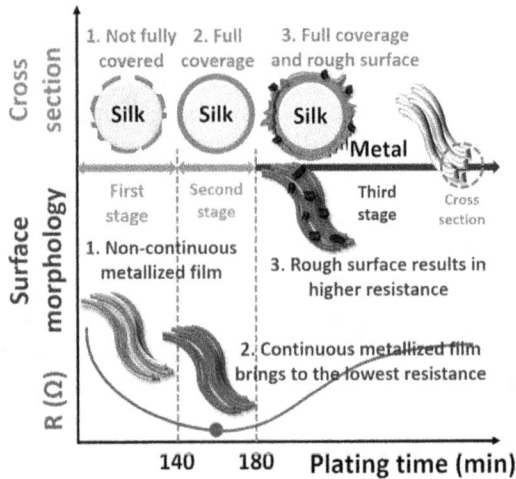

Fig. 6.56. Relationships between the cross-section, surface morphology, and electrical resistance in different stages.

One of the key factors to trigger the allergy reaction is the ion released from the metal, and the Pt ion released from the silk-Pt composite would be the concern in this case. Therefore, examining the ion releasing rate in the SBF is a direct method to evaluate the biocompatibility of the silk-Pt composite. The Pt ion concentration in human body varies with race, environment, and diet. According to several literature based on different living conditions, the average Pt input from diet is 1.44 µg/day, and the Pt output through the faeces can be up to 1.79 µg/day [46-49]. Regarding evaluation of the ion releasing rate of the silk-Pt composite prepared in this study, only 2.73 µg of Pt was found in the SBF after 3 months of immersion at 37 °C, which is the human body temperature. The average amount of Pt released from the silk-Pt composite per day would be around 0.03 µg, which is negligible when compared with the daily Pt input. Moreover, based on the daily Pt output, no Pt would accumulate in human body. This result provides a direct evidence that the silk-Pt composite material fabricated in the study is feasible for applications in implantable medical and wearable devices.

226

6.3.2.2. Pd(acac)$_2$ Sc-CO$_2$ Catalyzation and CONV Metallization

Identical reaction cell, high pressure system, and catalyzation conditions were used as the catalyzation cell. Silk textile with dimension at 2×4 cm^2 and Pd(acac)$_2$ were used as the substrate and the catalyst, respectively. Same Pt metallization solution was used in the section. Twenty-five milligrams of Pd(acac)$_2$ was introduced into the cell to maintain the catalyst concentration at the saturation point throughout the catalyzation process. Similar Pt metallization was conducted in this section. In order to confirm the reduction of Pd(acac)$_2$ to Pd metal, a piece of sc-CO$_2$ promoted catalyzation treated silk was immersed in a 2 vol. % reduction agent solution (pH = 12) at 70 ± 1 °C under ambient pressure with agitation for 30 min. The surface morphology, cross-section, thickness, compositions, and phases were observed by OM, SEM, EDX, and XRD respectively. The corrosion resistance was measured by polarization measurements in both 3.5 wt.% NaCl and a SBF solution. The 3.5 wt.% NaCl solution was used to simulate human sweat [50]. These two solutions were thus chosen for the in vitro and in vivo simulation, respectively. The potentiodynamic polarization curves were measured by using a Potentiostat/Galvanostat. The silk-Pt composite with 120 min of the metallization time, a Pt plate, and an Ag/AgCl reference electrode were used as the working electrode, the counter electrode, and the reference electrode, respectively. The total surface area of the working electrode, which is a piece of the 1×1 cm^2 silk-Pt composite, was calculated by considering the surface area of every bundle of the Pt metallized silk textile, and the value was 9.5 cm^2. The Pt counter electrode was cleaned ultrasonically in ethanol and pure water for 1 min, respectively, before the polarization measurements. The scan range started from -0.4 to $+0.6$ V vs Ag/AgCl at a scan rate at 1 mV/s. Prior to the potentiodynamic test, the three electrodes were immersed into the solution for 25 min to certify the stability. Similar immersion test and ICP measurements were carried out in this section. Electrical resistance and adhesive test of the composite materials were measure with similar method in the previous sections. The electrical resistance measurement was carried out ten times to ensure the reliability of the values reported and construct the error bar.

Fig. 6.57 shows the OM images of the non-treated pure silk, catalyzed silk, and catalyzed silk with reduction treatment. Transparent thread bundles shown in Fig. 6.57 indicate the silk textile; light yellow flakes shown in Fig. 6.57 correspond to Pd(acac)$_2$ catalysts, and the small particles with metallic luster shown in Fig. 6.57 are the reduced Pd metals. As shown in Fig. 6.57, the catalyst is successfully settled on the substrate without damaging the silk structure. After the reduction, small Pd particles finely remained on the substrate shown in Fig. 6.57.

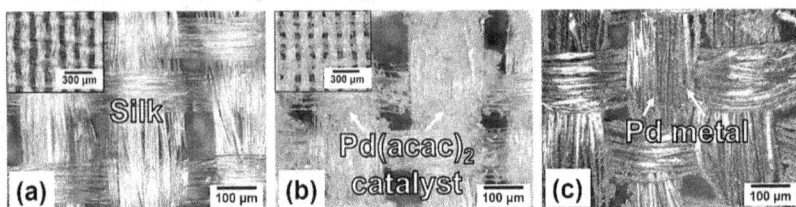

Fig. 6.57. OM images of (a) non-treated pure silk, (b) catalyzed silk, and (c) catalyzed silk after reduction.

XRD patterns of the (a) pure silk, (b) catalyzed silk, and (c) catalyzed silk with reduction treatment are shown in Fig. 6.58. The pure silk shows one broad peak at around $2\theta = 20°$ in Fig. 6.58. On the other hand, after the catalyzation, all the XRD characteristic peaks can be indexed to the contribution from Pd(acac)$_2$ [51, 52], shown in Fig. 6.58b. Fig. 6.58c shows the XRD pattern of the reduced catalyst on silk. Four diffraction peaks at $2\theta = 40.1°$, $46.7°$, $68.1°$, and $82.1°$ can be indexed to (111), (200), (220), and (311) planes of the FCC structure of palladium (JCPDS #89-4897). The broad peak observed at $2\theta = 20°$ in the XRD pattern indicates the amorphous structure of silk textile as shown in Fig. 6.58a. With the introduction of sc-CO$_2$, its nonpolar and low viscosity bring the catalyst into the substrate without damaging the substrate structure. The OM image (Fig. 6.57b) along with the characteristic peaks in Fig. 6.58b reveal that the catalyst was successfully settled on the substrate. Due to the high intensity from the Pd(acac)$_2$ catalysts, the broad peak from silk was suppressed and was not observed in Fig. 6.58b. According to Vrubel et al. [53] and Tanaka et al. [54], the reduction reaction of Pd(acac)$_2$ is believed to be the reactions listed in the following:

$$Pd(C_5H_7O_2)_2 + 2N_2H_4 = Pd^{2+} + 2C_5H_8N_2 + 2OH^- + 2H_2O,$$

where $Pd(C_5H_7O_2)_2 = Pd(acac)_2$

$$2Pd^{2+} + N_2H_4 + 4OH^- = 2Pd + N_2 + 4H_2O.$$

The broad peak at around $2\theta = 20°$ in Fig. 6.58c is attributed to the amorphous silk substrate, and the sharp peaks indicate successful reduction of the $Pd(acac)_2$ to form Pd metal. The results of Figs. 6.58a-c are in good agreement with Figs. 6.57a-c. The reduced catalysts can be used as the active sites for the deposition of Pt metals in the metallization step.

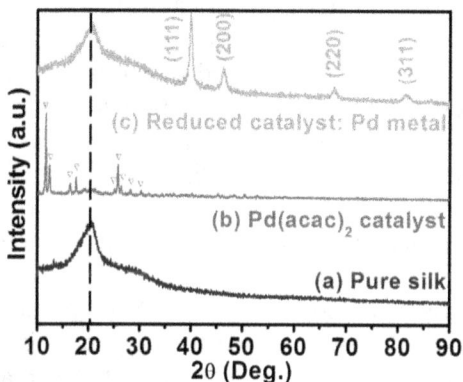

Fig. 6.58. X-ray diffraction patterns of (a) non-treated pure silk, (b) silk with sc-CO_2 catalyzation treatment, and (c) catalyzed silk with the reduction treatment.

The three SEM images shown in Fig. 6.59 demonstrate the trend of coverage of the Pt layer on the silk with various metallization times. Fig. 6.59a shows the SEM image of the sample with 80 min of the metallization time, where the bright phase corresponds to un-metallized silk while the gray phase indicates Pt metal. Since the electron cannot be conveyed on the nonconductive silk, thus the electrons accumulated on the un-metallized silk would show the charging effect. At 80 min of the metallization time, some parts of silk textile were un-metallized (pointed by arrows), showing a noncontinuous Pt layer and a smooth surface on the metallized part. Fig. 6.59b shows the morphology of the sample with 120 min of the metallization time; a continuous Pt layer was constructed showing gray color thoroughly, and the surface remained smooth. When the metallization time was extended to 180 min (Fig. 6.59c), the silk substrate was also completely covered. However, undesired Pt clusters accumulated on the surface, and the metallized surface became rough (pointed by arrows).

Based on the coverage, surface roughness, and electrical resistance, the Pt metallization of silk substrates is classified into three stages. The partially metallized silk with a discontinuous Pt layer at 80 min of the metallization time is classified to the first stage (Fig. 6.59a). The partly metallized surface showed smooth morphology since the metallization is still in the early stage. In this stage, the electrical resistance remained high due to the discontinuous Pt layer. At 120 min of the metallization time (Fig. 6.59b), the silk substrate was fully covered with Pt, which is classified into the second stage. Electrical resistance was lowered since a continuous Pt layer was constructed. The metallized surface was smooth, and there were not many redundant clusters on the Pt layer. Hence, the lowest electrical resistance at 55 mΩ was built up in the second stage at the optimal metallization time of 120 min. At 180 min of the metallization time (Fig. 6.59c), the silk was still fully covered; however, undesired Pt clusters and the rough metallization resulted in the elevated electrical resistance. The rough Pt deposition on the substrate surface increased unnecessary Joule heating and caused formation of a relative loose structure. A proportional relationship between the electrical resistance and the roughness is indicated in the literature [42]. Therefore, coverage of the Pt layer and the surface roughness both have significant influence on the electrical resistance.

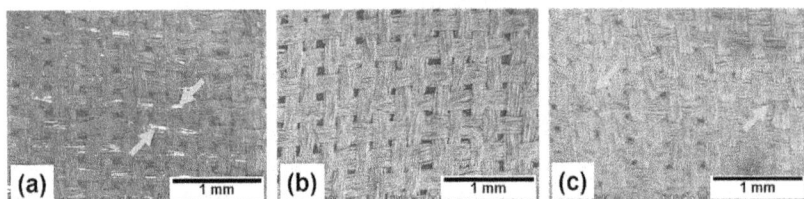

Fig. 6.59. Pt metallization trend demonstrates by SEM images at (a) 80 min, (b) 120 min, and (c) 180 min of the metallization time.

In the same way, elemental mapping of the Pt metallized silk (Fig. 6.60) shows the critical time demarcation of full coverage, which lies between (a) 80 and (b) 100 min of the metallization time. The least time needed to construct full coverage is defined as the minimum time, which is 100 min. The composition results show a pure Pt layer with merely trace amount of impurity.

Fig. 6.61 shows the XRD pattern after the Pt metallization. Five diffraction peaks located at $2\theta = 40.2°$, $46.8°$, $68.4°$, $82.4°$, and $86.9°$ can be indexed to the (111), (200), (220), (311), and (222) planes of the FCC structure of platinum (JCPDS #87-0647).

Fig. 6.60. Elemental mapping of metallized silks at (a) 80 min and (b) 100 min of the metallization time.

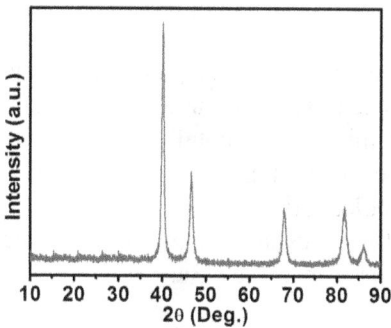

Fig. 6.61. X-ray diffraction pattern of Pt metallized silk substrate.

The metallization time dependence of the Pt layer thickness is presented in Fig. 6.62. A positive correlation between the metallization time and the thickness is shown. A reliable thickness value could not be measured for the samples with a metallization time less than the minimum time at

100 min due to the poor electrical conductivity. Pt layer thickness at 0.90 μm was obtained with the longest metallization time at 180 min, for which the calculated average growth rate is 1.8 nm/min.

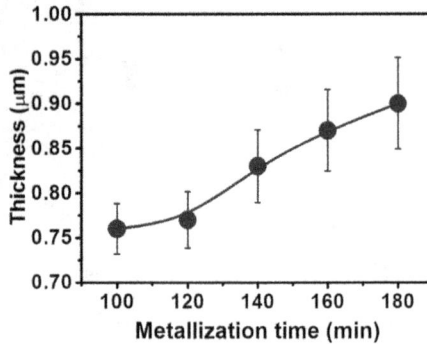

Fig. 6.62. Pt layer thickness as a function of the metallization time.

Releasing of the metal ions is an essential factor to provoke the allergic reactions; [14] therefore, investigating the released ion concentration in the SBF is a direct way to evaluate the biocompatibility of the silk-Pt composite. After three months of the immersion at human body temperature (37 °C) in the SBF, only 0.011 μg/day and 0.018 μg/day of Pd and Pt were observed, respectively.

Fig. 6.63 shows the results of the polarization measurements of the silk-Pt composite materials before and after the adhesive test. The silk substrate with 120 min of the metallization time was chosen for the corrosion resistance evaluation. No significant difference was found among the three polarization curves, indicating excellent corrosion resistance of the Pt layer even after the second adhesive test. The E_{corr} and I_{corr} are summarized in Table 6.3a.

For simulating the condition in the human body, similar polarization measurements were carried out in the SBF. Fig. 6.64 shows the polarization curves of the silk-Pt composite materials before and after the adhesive test. In the same way, the metallization time at 120 min was chosen for its best suitability to this application. The corrosion resistance also persisted after the second adhesive test, and no significant difference within these three polarization curves was resistance as a function of the metallization time before and after the adhesive test.

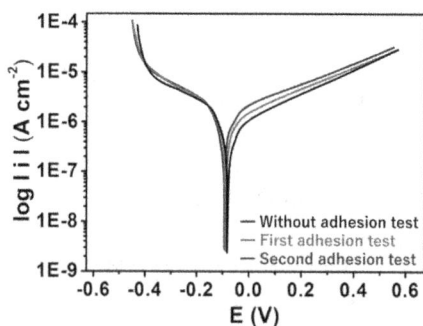

Fig. 6.63. Polarization measurements of the silk-Pt composites before and after the adhesive test in 3.5 wt.% NaCl solution.

Table 6.3. Corrosion potentials and corrosion currents of the silk-Pt composites in (a) 3.5 wt.% NaCl and (b) SBF solutions.

(a) Samples (in 3.5 wt.% NaCl)	E_{corr} (V)	I_{corr} (A/cm^2)
As-deposited	-0.090	3.58×10^{-7}
Adhesive test for once	-0.092	3.50×10^{-7}
Adhesive test for twice	-0.093	3.56×10^{-7}
(b) Samples (in SBF)	E_{corr} (V)	I_{corr} (A/cm^2)
As-deposited	0.015	2.16×10^{-7}
Adhesive test for once	0.018	2.02×10^{-7}
Adhesive test for twice	0.019	2.02×10^{-7}

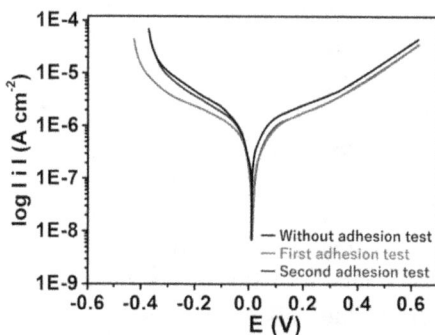

Fig. 6.64. Polarization measurements of the silk-Pt composites before and after the adhesive test in the SBF.

233

Again, a reliable electrical resistance cannot be obtained owing to the poor conductivity when the metallization time is less than 80 min. The electrical resistance was high in the early stage of the metallization and then decreased with an increasing in the metallization time. After reaching a minimum point at 120 min, the electrical resistance gradually increased and stabilized. The electrical resistances after the adhesive test are shown in Fig. 6.65. The adhesive tests showed serious impact on the samples in the early stage. On the contrary, the electrical resistances persist after the second adhesive test in the last two stages. In addition, there is no distinct difference between the first and the second adhesive test. The stages can be classified into three according to the morphology and the electrical resistance, which will be discussed in the following paragraph.

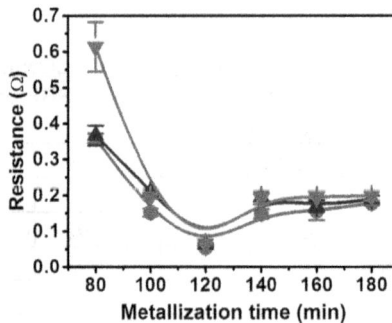

Fig. 6.65. Electrical resistance curve as a function of metallization time of the silk-Pt composites before and after the adhesive test.

3.5 wt.% NaCl solution, high corrosion resistance of the silk-Pt is shown in the polarization measurements in Fig. 6.63a. On the other hand, the I_{corr} of the composite materials in the SBF are all lower than those in 3.5 wt.% NaCl solution since 3.5 wt.% NaCl solution contains more Cl^-, and Cl^- has significant impact on the corrosion resistance of metals [55]. The E_{corr} of silk-Pt composite in the SBF is also more positive than those in 3.5 wt.% NaCl solution, indicating that the Pt layer is more inert in the SBF solution. Despite the fact that a thin Pt layer was electroless plated on the nonconductive substrates, the corrosion rates are comparable to those of bulk Pt metal, which indicates the well deposition of the Pt layer on the silk. Overall, the silk-Pt composites show high corrosion resistances in the two solutions. These results suggest that the silkPt composite is not merely practical in wearable devices but also in implantable devices.

The metal ion concentration in the human body varies with race, environment, and diet. According to several literature reports based on different living conditions, the average Pt input from diet is 1.44 µg/day, and Pt output through the faeces can be up to 1.79 µg/day [48, 56-58]. On the other hand, the average Pd intake is 2 µg/day, and the Pd output can be up to 2.5 µg/day [59]. From the immersion test, only 0.018 µg Pt/day and 0.011 µg Pd/day were observed in the SBF solution. Both values are far less than the intake and have no chance to accumulate in the human body according to the literature and the ICP results. In addition, the Pt ion releasing rate of the silk-Pt composite prepared in this study is lower than those catalyzed by the $Pt(acac)_2$. Since the samples catalyzed by $Pd(acac)_2$ show relatively smooth deposition and less defects, they are thus corroded mildly in the immersion tests while compared to those catalyzed by $Pt(acac)_2$.

The electroless plating system catalyzed by $Pd(acac)_2$ demonstrates a more efficient process and improved material properties compared to those of the system catalyzed by $Pt(acac)_2$. The optimal metallization time needed to reach the lowest electrical resistance is at 120 min, which shows 25 % reduction in the time when compared to the study using $Pt(acac)_2$ as the catalyst. The improved efficiency is suggested to be attributed to the high reduction efficiency of $Pd(acac)_2$ to Pd metal. According to the literature [60], $Pd(acac)_2$ owns a lower decomposition temperature than that of $Pt(acac)_2$; it thus can be decomposed or reduced to the metal state more efficiently. After formation of the activation sites and gradual deposition of the Pt layer on the silk surface, the catalysts would be fully covered by the Pt and have no significant influence on the succeeding Pt deposition rate. It is worthy of mention that the silk-Pt composites fabricated in this study showed electrical resistances 36 % lower than those of the literature. When comparing with the reduction of $Pd(acac)_2$ to Pd metals, reduction of $Pt(acac)_2$ to Pt metals is less efficient and takes a longer time. The non-synchronized $Pt(acac)_2$ reduction leads to uneven growth of the Pt activation sites and the Pt metallization later on. The irregular sizes of Pt activation sites make the succeeding Pt deposition grow roughly and thus show the high electrical resistance. By using $Pd(acac)_2$, nucleation and growth of the activation site would be uniform and lead to even growth of the Pt layer. Hence, the electrical resistance could be lowered.

The levels of damage severity from the adhesive test were evaluated by both corrosion and electrical resistance measurements. Generally, defects and cracks would be formed after the adhesive test. Then the corrosion test solution can penetrate into the defects and cracks to give an increase in the current during the polarization measurement. Since the surface area is assumed to be constant, the current density is thus expected to be increased after formation of the defects and cracks. However, no significant difference in the I_{corr} was found after the adhesive test, indicating that the samples stay intact after the adhesive test and hence showing the high adhesive property of the Pt layer on the silk. In the electrical resistance test, the electrical resistance is expected to be elevated if the adhesive test successfully introduces defects into the Pt layer due to an increase in surface roughness of the Pt layer. The results shown in Fig. 6.65 indicate great impacts in the first stage; on the contrary, the electrical resistance persists in the final two stages. In the first stage, the Pt layer is not continuous and the edge of the Pt layer is exposed to the tape during the adhesive tests. The Pt edge would be peeled off more easily than the surface that is fully metallized since it is free from support in the edge. The silk metallized with the optimized time shows perseverance in the second adhesive test, indicating its feasibility in the practice of applications even when under adverse conditions.

The influence of morphology on the electrical resistance in different stages is illustrated in Fig. 6.66.

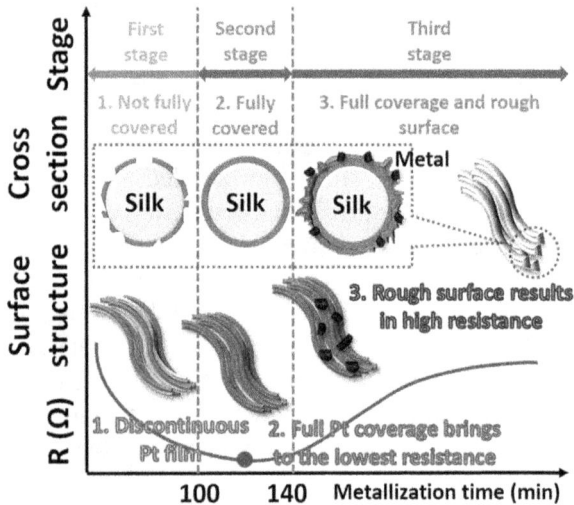

Fig. 6.66. Relationship between metallization morphology and electrical resistance in three different stages.

In the first stage, the silk was not fully metallized, showing a noncontinuous Pt layer. It thus gave the highest electrical resistance. When the metallization time was prolonged to the second stage, full coverage was constructed, showing a continuous Pt layer. Along with the smooth surface, the electrical resistance reached its minimum point at 120 min in the second stage. However, the rough surface emerged as the metallization time extended; the electrical conductivity was thus deteriorated in the third stage.

6.4. Conclusions

Ni-P and Pt were successfully metallized on the polyimide, nylon 6,6, and silk textiles. Sc-CO_2-assisted catalyzation shows better adhesive properties between metallization layer and polymers. Polymers were not damaged after sc-CO_2-assisted catalyzation, while it was damaged seriously with CONV catalyzation due to the acidic catalyzation solution. Sc-CO_2-assisted metallization shows smooth metallization layer and no obvious roughness increasing was found in this technique. Thickness of metallization layer increased with increasing metallization time durations. All the diffraction peaks corresponded to Pt(acac)$_2$ catalyst, Pd(acac)$_2$ catalyst, Ni-P, and Pt metal well in different stages. Metallized polymers show the lowest electrical conductivity with the optimized metallization condition. Three stages were found in the metallization according to the surface morphology and electrical resistance. First stage shows non-fully covered polymer, second stage shows fully covered polymer and smooth surface, and third stage shows fully covered polymer and rough surface. The roughened metallization surface brings the deteriorated electrical resistance. Metallized polymer shows high peel-off resistance in the adhesive test. The electrical resistance remains almost the same after twice adhesive test in second and third stages with full coverage. While the composite materials in the first stage was damaged significantly in the peeling test. Ni-P metallization shows high corrosion resistance in 3.5 wt.% NaCl solution. On the other hand, Pt metallized polymer shows high biocompatibility and high corrosion resistance in both in vitro and in vivo solutions.

Acknowledgements

This work is supported by Japan Society for the Promotion of Science (17J07395), the Grant-in-Aid for Scientific Research (B) (JSPS

KAKENHI Grant Number 26282013), CREST Project operated by the Japan Science and Technology Agency (JST) (#14531864).

Reference

[1]. J. Gozalvez, 5G tests and demonstrations [mobile radio], *IEEE Vehicular Technology Magazine*, Vol. 10, 2015, pp. 16-25.

[2]. A. J. Bandodkar, I. Jeerapan, J. Wang, Wearable chemical sensors: Present challenges and future prospects, *ACS Sensors*, Vol. 1, 2016, pp. 464-482.

[3]. Y.-S. Wang, S.-M. Li, S.-T. Hsiao, W.-H. Liao, P.-H. Chen, S.-Y. Yang, H.-W. Tien, C.-C. M. Ma, C.-C. Hu, Integration of tailored reduced graphene oxide nanosheets and electrospun polyamide-66 nanofabrics for a flexible supercapacitor with high-volume-and high-area-specific capacitance, *Carbon*, Vol. 73, 2014, pp. 87-98.

[4]. G. O. Mallory, J. B. Hajdu, Electroless Plating: Fundamentals and Applications, *William Andrew*, 1990.

[5]. Y. Shacham-Diamand, A. Inberg, Y. Sverdlov, V. Bogush, N. Croitoru, H. Moscovich, A. Freeman, Electroless processes for micro-and nanoelectronics, *Electrochimica Acta*, Vol. 48, 2003, pp. 2987-2996.

[6]. H. Adachi, K. Taki, S. Nagamine, A. Yusa, M. Ohshima, Supercritical carbon dioxide assisted electroless plating on thermoplastic polymers, *The Journal of Supercritical Fluids*, Vol. 49, 2009, pp. 265-270.

[7]. G. Brunner, Gas Extraction: An Introduction to Fundamentals of Supercritical Fluids and the Application to Separation Processes, *Springer Science & Business Media*, 2013.

[8]. S. P. Sawan, S. P. Sawan, Supercritical Fluid Cleaning: Fundamentals, Technology and Applications, *Elsevier*, 1998.

[9]. A. Clifford, T. Clifford, Fundamentals of Supercritical Fluids, *Oxford University Press*, 1999.

[10]. M. Sano, Y. Tahara, C.-Y. Chen, T.-F. M. Chang, T. Hashimoto, H. Kurosu, T. Sato, M. Sone, Application of supercritical carbon dioxide in catalyzation and Ni-P electroless plating of nylon 6,6 textile, *Surface and Coatings Technology*, Vol. 302, 2016, pp. 336-343.

[11]. T.-F. M. Chang, T. Tasaki, C. Ishiyama, M. Sone, Defect-free nickel micropillars fabricated at a high current density by application of a supercritical carbon dioxide emulsion, *Industrial & Engineering Chemistry Research*, Vol. 50, 2011, pp. 8080-8085.

[12]. X. L. Huang, D. Xu, S. Yuan, D. L. Ma, S. Wang, H. Y. Zheng, X. B. Zhang, Dendritic Ni-P-coated melamine foam for a lightweight, low-cost, and amphipathic three-dimensional current collector for binder-free electrodes, *Advanced Materials*, Vol. 26, 2014, pp. 7264-7270.

[13]. A. Cowley, B. Woodward, A healthy future: platinum in medical applications, *Platinum Metals Review*, Vol. 55, 2011, pp. 98-107.

[14]. J. C. Wataha, Principles of biocompatibility for dental practitioners, *The Journal of Prosthetic Dentistry*, Vol. 86, 2001, pp. 203-209.

[15]. A. Oyane, H. M. Kim, T. Furuya, T. Kokubo, T. Miyazaki, T. Nakamura, Preparation and assessment of revised simulated body fluids, *Journal of Biomedical Materials Research Part A*, Vol. 65, 2003, pp. 188-195.

[16]. R. Bella, P. Cassagnau, F. Fenouillot, L. Falk, C. Lacoste, Diffusion of liquids in molten polymers: Mutual diffusion coefficient dependence on liquid miscibility and polymer molar mass, *Polymer*, Vol. 47, 2006, pp. 5080-5089.

[17]. Y. Sato, T. Takikawa, A. Sorakubo, S. Takishima, H. Masuoka, M. Imaizumi, Solubility and diffusion coefficient of carbon dioxide in biodegradable polymers, *Industrial & Engineering Chemistry Research*, Vol. 39, 2000, pp. 4813-4819.

[18]. P. Lau, C. Wong, L. Chan, Improving electroless Cu via filling with optimized Pd activation, *Applied Surface Science*, Vol. 253, 2006, pp. 2357-2361.

[19]. H. Wakabayashi, N. Sato, M. Sone, Y. Takada, H. Yan, K. Abe, K. Mizumoto, S. Ichihara, S. Miyata, Nano-grain structure of nickel films prepared by emulsion plating using dense carbon dioxide, *Surface and Coatings Technology*, Vol. 190, 2005, pp. 200-205.

[20]. H. Wakayama, Y. Fukushima, Porous platinum fibers synthesized using supercritical fluid, *Chemical Communications*, Vol. 0, Issue 4, 1999, pp. 391-392.

[21]. G. S. Paul, Diffusion in Solids, *McGraw-Hill Publishing*, 1963.

[22]. Y. Sato, T. Takikawa, S. Takishima, H. Masuoka, Solubilities and diffusion coefficients of carbon dioxide in poly (vinyl acetate) and polystyrene, *The Journal of Supercritical Fluids*, Vol. 19, 2001, pp. 187-198.

[23]. J. Patterson, C. N. Dheasuna, J. Barrett, T. Spalding, M. O'Reilly, X. Jiang, G. Crean, Electroless copper metallisation of titanium nitride, *Applied Surface Science*, Vol. 91, 1995, pp. 124-128.

[24]. J. Patterson, M. O'Reilly, G. Crean, J. Barrett, Selective electroless copper metallization on a titanium nitride barrier layer, *Microelectronic Engineering*, Vol. 33, 1997, pp. 65-73.

[25]. S. Nakahara, Y. Okinaka, Microstructure and Mechanical Properties of electroless copper deposits, *Annual Review of Materials Science*, Vol. 21, 1991, pp. 93-129.

[26]. I. Kikic, F. Vecchione, Supercritical impregnation of polymers, *Current Opinion in Solid State and Materials Science*, Vol. 7, 2003, pp. 399-405.

[27]. L. N. Nikitin, M. O. Gallyamov, R. A. Vinokur, A. Y. Nikolaec, E. E. Said-Galiyev, A. R. Khokhlov, H. T. Jespersen, K. Schaumburg, Swelling and impregnation of polystyrene using supercritical carbon dioxide, *The Journal of Supercritical Fluids*, Vol. 26, 2003, pp. 263-273.

[28]. S. Yoda, A. Hasegawa, H. Suda, Y. Uchimaru, K. Haraya, T. Tsuji, K. Otake, Preparation of a platinum and palladium/polyimide

nanocomposite film as a precursor of metal-doped carbon molecular sieve membrane via supercritical impregnation, *Chemistry of Materials*, Vol. 16, 2004, pp. 2363-2368.

[29]. T. Miyamura, J. Koike, The effects of Cr oxidation and polyimide degradation on interface adhesion strength in Cu/Cr/polyimide flexible films, *Materials Science and Engineering: A*, Vol. 445, 2007, pp. 620-624.

[30]. S. H. Kim, S. H. Cho, N.-E. Lee, H. M. Kim, Y. W. Nam, Y.-H. Kim, Adhesion properties of Cu/Cr films on polyimide substrate treated by dielectric barrier discharge plasma, *Surface and Coatings Technology*, Vol. 193, 2005, pp. 101-106.

[31]. R. M. McDonogh, C. J. D. Fell, A. G. Fane, Characteristics of membranes formed by acid dissolution of polyamides, *Journal of Membrane Science*, Vol. 31, 1987, pp. 321-336.

[32]. A. Małecki, A. Micek-Ilnicka, Electroless nickel plating from acid bath, *Surface and Coatings Technology*, Vol. 123, 2000, pp. 72-77.

[33]. U. K. Fatema, Y. Gotoh, A new electroless Ni plating procedure of iodine-treated aramid fiber, *Journal of Coatings Technology and Research*, Vol. 10, 2013, pp. 415-425.

[34]. H. Yoshida, M. Sone, A. Mizushima, K. Abe, X. T. Tao, S. Ichihara, S. Miyata, Electroplating of nanostructured nickel in emulsion of supercritical carbon dioxide in electrolyte solution, *Chemistry Letters*, Vol. 31, 2002, pp. 1086-1087.

[35]. T.-F. M. Chang, M. Sone, A. Shibata, C. Ishiyama, Y. Higo, Bright nickel film deposited by supercritical carbon dioxide emulsion using additive-free Watts bath, *Electrochimica Acta*, Vol. 55, 2010, pp. 6469-6475.

[36]. T.-F. M. Chang, M. Sone, Function and mechanism of supercritical carbon dioxide emulsified electrolyte in nickel electroplating reaction, *Surface and Coatings Technology*, Vol. 205, 2011, pp. 3890-3899.

[37]. H. Uchiyama, M. Sone, A. Shibata, Y. Higo, Effects of CO_2 on Ni-P electroless plating in an emulsion of supercritical CO_2, *Journal of The Electrochemical Society*, Vol. 157, 2010, pp. D550-D552.

[38]. H. Uchiyama, T. Endo, M. Sone, Direct observation of nodule growth on electroless Ni-P deposition in supercritical CO_2 emulsion, *Journal of the Electrochemical Society*, Vol. 159, 2011, pp. D114-D118.

[39]. Y. S. Cheng, K. L. Yeung, Effects of electroless plating chemistry on the synthesis of palladium membranes, *Journal of Membrane Science*, Vol. 182, 2001, pp. 195-203.

[40]. L. Abrantes, J. Correia, On the Mechanism of Electroless Ni-P Plating, *Journal of the Electrochemical Society*, Vol. 141, 1994, pp. 2356-2360.

[41]. W. Tang, K. Xu, P. Wang, X. Li, Surface roughness and resistivity of Au film on Si-(111) substrate, *Microelectronic Engineering*, Vol. 66, 2003, pp. 445-450.

[42]. V. Timoshevskii, Y. Ke, H. Guo, D. Gall, The influence of surface roughness on electrical conductance of thin Cu films: an ab initio study, *Journal of Applied Physics*, Vol. 103, 2008, 113705.

[43]. C. J. Lin, K. C. Chen, J. L. He, The cavitation erosion behavior of electroless Ni-P-SiC composite coating, *Wear*, Vol. 261, 2006, pp. 1390-1396.

[44]. J. N. Balaraju, Kalavati, K. S. Rajam, Influence of particle size on the microstructure, hardness and corrosion resistance of electroless Ni-P-Al$_2$O$_3$ composite coatings, *Surface and Coatings Technology*, Vol. 200, 2006, pp. 3933-3941.

[45]. A. Bai, P.-Y. Chuang, C.-C. Hu, The corrosion behavior of Ni-P deposits with high phosphorous contents in brine media, *Materials Chemistry and Physics*, Vol. 82, 2003, pp. 93-100.

[46]. G. T. Vaughan, T. M. Florence, Platinum in the human diet, blood, hair and excreta, *Science of The Total Environment*, Vol. 111, 1992, pp. 47-58.

[47]. C. M. G. van den Berg, G. S. Jacinto, The determination of platinum in sea water by adsorptive cathodic stripping voltammetry, *Analytica Chimica Acta*, Vol. 211, 1988, pp. 129-139.

[48]. E. D. Goldberg, V. Hodge, P. Kay, M. Stallard, M. Koide, Some comparative marine chemistries of platinum and iridium, *Applied Geochemistry*, Vol. 1, 1986, pp. 227-232.

[49]. E. D. Goldberg, M. Koide, Understanding the marine chemistries of the platinum group metals, *Marine Chemistry*, Vol. 30, 1990, pp. 249-257.

[50]. Y. Song, D. Shan, E. Han, Corrosion behaviors of electroless plating Ni-P coatings deposited on magnesium alloys in artificial sweat solution, *Electrochimica Acta*, Vol. 53, 2007, pp. 2009-2015.

[51]. D. Esken, X. Zhang, O. I. Lebedev, F. Schröder, R. A. Fischer, Pd@MOF-5: limitations of gas-phase infiltration and solution impregnation of [Zn$_4$O(bdc)$_3$](MOF-5) with metal-organic palladium precursors for loading with Pd nanoparticles, *Journal of Materials Chemistry*, Vol. 19, 2009, pp. 1314-1319.

[52]. Y. Wang, H. Zheng, Z. Li, K. Xie, Investigation of the interaction between Cu(acac)$_2$ and NH$_4$Y in the preparation of chlorine-free CuY catalysts for the oxidative carbonylation of methanol to a fuel additive, *RSC Advances*, Vol. 5, 2015, pp. 102323-102331.

[53]. H. Vrubel, V. H. C. Verzenhassi, S. Nakagaki, F. S. Nunes, Catalytic reduction of hydrazine to ammonia by a high-oxidation state molybdenum complex, *Inorganic Chemistry Communications*, Vol. 11, 2008, pp. 1040-1043.

[54]. D. A. P. Tanaka, M. A. L. Tanco, S.-I. Niwa, Y. Wakui, F. Mizukami, T. Namba, T. M. Suzuki, Preparation of palladium and silver alloy membrane on a porous α-alumina tube via simultaneous electroless plating, *Journal of Membrane Science*, Vol. 247, 2005, pp. 21-27.

[55]. F.-Y. Ma, Corrosive Effects of Chlorides on Metals, Pitting Corrosion, *InTech*, 2012.

[56]. G. Vaughan, T. Florence, Platinum in the human diet, blood, hair and excreta, *Science of the Total Environment*, Vol. 111, 1992, pp. 47-58.

[57]. C. M. van den Berg, G. S. Jacinto, The determination of platinum in sea water by adsorptive cathodic stripping voltammetry, *Analytica Chimica Acta*, Vol. 211, 1988, pp. 129-139.

[58]. E. D. Goldberg, M. Koide, Understanding the marine chemistries of the platinum group metals, *Marine Chemistry*, Vol. 30, 1990, pp. 249-257.

[59]. I. Iavicoli, B. Bocca, L. Fontana, S. Caimi, A. Bergamaschi, A. Alimonti, Distribution and elimination of palladium in rats after 90-day oral administration, *Toxicology and Industrial Health*, Vol. 26, 2010, pp. 183-189.

[60]. H. Koubu, Y. Hayashi, J. Fukushima, H. Takizawa, I. Narita, S. Yoshioka, Synthesis of noble metal-doped Cu nanoparticles by ultrasonication, *Materials Transactions*, Vol. 54, 2013, pp. 1496-1501.

Index

2D material, 149

A

acoustic, 79, 93, 94, 99, 101, 111, 119, 120, 122, 123, 127, 128, 132
acoustical sensing, 68
activated partial thromboplastin time, 91
adhesion, 82, 85-87, 89, 92, 101, 119, 121, 123, 124, 126
adhesive, 167, 209, 214-217, 236
adhesive test, 167, 209, 214-217
Adsorption, 95
air quality control, 78, 132
allergic reactions, 232
allergy, 167, 226
amorphous, 170, 210, 211, 222, 228, 229
amperometric H_2O_2 detection, 152
analytical tool, 78, 80, 112
anchoring effect, 198, 200, 204, 220
anticoagulation, 89, 125
Antithrombotic, 89
artificial membranes, 67, 68, 81, 82
autocatalytic, 181

B

Barton, 104, 131
biocompatibility, 166, 167, 226, 232
bioelectronics, 67, 106, 110
biofouling, 92, 111, 126
bioimpedance, 67, 70, 71, 75-77, 110, 112-117
bioimprinted polymers, 82
Bioimprinting, 83
biological recognition, 17
biomimetics, 82, 84

biosensors, 16, 71, 81, 82, 84, 100, 102, 104, 105, 108, 120, 121, 127-129, 132
bis (2,4-pentandionato)-platinum, 167
blood
    cells, 67, 68, 77, 85, 87, 89, 92, 111, 124
    clotting., 68
    purification, 67, 89, 110
body fluid, 67, 70, 71, 75-77, 113
Bovine serum albumin, 102

C

cancer
    biomarker, 147
    cell, 157, 161
    diagnosis, 147
catalyst, 166, 169-171, 173, 174, 176, 178-182, 185, 193, 195, 197-200, 204, 205, 207, 209-211, 215, 221-223, 227-229, 235
catalyzation, 166-190, 192-208, 210, 215, 218-222, 227-229
chemical
    plating, 165
    sensor, 133
Clark oxygen electrode, 16
clinical diagnosis, 15
clinically relevant molecules, 15
clotting, 68, 89, 92, 124, 125
coagulation, 68, 89, 90-92, 111, 124-126
colorimetric
    detection, 133-135, 139
    sensor, 136, 137, 139, 142
composite, 166, 167, 209, 212-218, 221, 223, 225-227, 232-235
composite semiconductors, 24
conduction and valence band, 22

conformal deposition, 166
conjugated polymers
  polydiacetylene, 134
conventional, 166
Corrosion resistance, 167, 209
coupling, 80, 99, 123
Crit-Line monitor, 77
cyanobacteria, 107
Cyclic voltammetry, 153

D

dialysis, 66-69, 76, 77, 80, 92, 103,
  107, 108, 111, 112, 115, 116,
  124, 126, 130
dielectric dispersion, 67, 71, 78,
  114
dielectric spectroscopy, 70, 87,
  116, 117
diffusion coefficient, 171, 179, 189,
  190, 192, 195, 196, 199
Direct Charge Transfer, 30
DNA enzymes, 82, 84
DNAzyme, 104, 123
dry weight, 69, 111, 114, 115

E

electrical
  cell substrate impedance sensing,
    67, 68, 71
  conductivity, 167, 216, 232, 237
  resistance, 70, 74
electrochemical, 182, 199
electrochemical biosensors, 148
electroimpedance tomography, 76
electroless plating, 165-167, 195,
  196, 225, 235
electrospinning, 138, 139, 141
electrospun nanofibers, 134, 140
environment, 82, 86, 93, 108
environmental, 79, 104
Enzymatic Reaction, 31
enzyme based biosensors, 149
extracellular water, 67, 75

F

fibrin, 68, 90, 91
flexibility, 166
flexible fabrics, 165
Formation of Composite
  Semiconductors, 24
Fouling, 127
Frank, 100, 120, 122, 125

G

Giaever, 101, 129
glass substrates, 134
gold nanoparticles, 134
graphene, 149

H

Hanai, 67, 71, 73, 75, 78, 109, 114,
  117
Hanai-Koizumi, 67, 73
Hardesty, 100, 120
heavy metal, 103, 104, 105, 107,
  111, 132
hematocrit, 77, 117
hemodialysis, 66-70, 80, 89, 92,
  103, 110, 112, 114-117, 124-
  127, 130
hydrogels, 136
hydrogen peroxide (H_2O_2), 147
hydrophobic, 169, 170, 174, 178-
  180, 185-187, 189, 197

I

immersion test, 227, 235
immunosensor, 108, 128
impedance spectroscopy, 67, 71,
  76, 100-102, 104, 110, 111, 130
impedimetric, 68, 70, 104, 105,
  106, 109, 111
implantable, 167, 226, 234
impregnation, 194, 195, 200, 204,
  221
Inorganic nanomaterials, 149
interdialytic weight gain, 76, 113

interfacial polarization, 67, 71, 78, 88, 114
interstitial fluid, 67, 76
intracellular water, 67, 113, 117

J

Janshoff, 100, 119-123, 129

K

Kanazawa, 100, 118-120, 122
Kapton, 167, 173, 176, 184, 187
Keese, 101, 129
Koizumi, 114, 117

L

label-free, 71, 108, 130-132
Layered doubled Hydroxides, 149
localized surface plasmon resonance (LSPR), 26

M

Matthie, 75, 113
Maxwell, 67, 71-73, 96, 97, 109, 114, 117
Maxwell-Wagner, 67, 71, 73, 109, 117
metabolite, 20
Metal
   alloys, 139
   Ion Doping, 24, 28
metallization, 166, 168, 169, 171, 173-176, 178, 179, 183-189, 191-227, 229-232, 234-237
metallized, 166, 168-171, 173-176, 178, 180, 183, 185-187, 189, 193, 194, 197, 199, 209, 211-213, 217, 221-225, 227, 229-231, 236, 237
micelles, 166
microcystin-LR, 108, 132
microcystins, 107
microfluidics, 108
microgravimetric, 68, 130

missing mass, 99
mixture models, 71, 73, 74
modeling, 67, 117
Moissl, 112, 114-117
molecular imprinting of polymers, 82
Molecularly imprinted polymers, 137
multifrequency, 67, 77, 116, 117
multifrequency bioimpedance spectroscopy, 67, 117

N

Nanofibers, 138
nanohybrids, 156
Nanohybrids, 150
nanoparticles, 82, 92, 104, 127, 131
nanostructures, 150
Newtonian bulk liquid, 98
N-formyl-methionyl-leucyl-phenylalanine, 155
Ni-P, 167-179, 182, 183, 185-190, 192-202, 204-218
non-enzymatic H_2O_2 detection, 149
non-invasive, 67, 69, 71, 77, 100
non-ionic surfactant, 184
Nylon 6,6, 166, 201-206, 208, 218-220
nylon-6, 139, 141

O

organic dye sensitization, 24, 26

P

personalized health care, 15
phase angle, 70, 101, 116
photoactive species, 17
photoelectrochemical (PEC) process, 16
plasma activated medium, 157
platelet aggregation, 91
Platinum, 218, 221
point of care diagnosis, 133
point-of-care, 15

point-of-care testing, 15
polarization, 209, 227, 232, 234, 236
polyethersulfone, 92
polyethylene, 92
polyimide, 166-186, 188-190, 192-200
polymer, 167, 169, 171, 173, 174, 176-178, 180, 182-186, 188, 190, 192-195, 197-202, 219-221
polyoxyethylene lauryl ether, 184
polysulfone, 87, 92, 125, 126
protein adsorption, 82, 90, 92, 93, 121, 122, 129
protein markers, 19, *See* also, *See* also
prothrombin time, 91
Pt, 166, 209, 218-227, 229-237

Q

quantum dots, 82, 104, 131
quartz crystal microbalance, 68, 78-81, 108, 111, 118-127, 129, 130, 132

R

reactive oxygen species, 21
real time monitoring, 68, 103
recombination rates, 23
red blood cells, 87
redox enzyme, 81, 106

S

sacrificial electron donor/acceptor, 18
Sauerbrey, 80, 94, 118
self-assembled monolayers, 81, 82
self-diffusivity, 166
semiconductor quantum dots, 22
sensing mechanisms of glucose oxidation, 21
sensitization approaches, 23
    biosensor, 133, 134, 135-137, 139-141

shapes of RBCs, 87
silk, 166, 208-211, 212-218, 221-223, 225-237
simulated body fluid, 167
soft biomaterials, 80
Solid State Colorimetric Biosensors, 133
Steric Hindrance, 33
Substrates, 134, 141
supercritical CO_2, 166
Supported membranes, 81, 120
surface
    acoustic waves, 93, 94
    immobilized peptides, 82
    roughness, 182, 193, 215, 224, 225, 230, 236
synergy effects, 35

T

textile, 166, 201-204, 218-223, 225, 227-229
theoretical background, 67, 68, 71, 76, 80
Thromboelastography, 91
thrombosis, 89
total body water, 67, 69, 113, 115-117
toxins, 83, 103, 107-111, 132

U

ultrafiltration, 66, 67, 115

V

vesicles, 81-84, 108, 122
viscoelastic, 68, 80, 84, 87, 93, 94, 96-101, 120, 122, 124, 127, 129
viscoelastic properties, 68, 127
viscoelasticity, 80, 91, 94, 99, 119, 128
Voigt, 96, 97, 122

W

water

quality, 68, 80, 102, 103, 111,
112, 130
toxins, 84, 86, 109
wearable device, 165, 166, 167,
202, 217, 226, 234
Wegener, 100, 119, 122, 123, 129

wettability, 171, 180, 185

X

Xitron, 71, 76
X-ray diffractometer, 168